Grade 1 · Unit 1

Inspire Science
All About Plants

McGraw Hill Education

mheducation.com/prek-12

Inspire Science

Explore Our Phenomenal World

Learning begins with curiosity. Inspire Science provides an in-depth, collaborative, and project-based learning experience designed to help you spark students' interest and empower them to ask more questions and think more critically. Through inquiry-based, hands-on investigations of real-world phenomena, your students will be able to answer more rigorous science questions with evidence, and generate innovative solutions to real-world problems.

Are you ready to inspire the next generation of innovators?

 ### Inspire Curiosity
Spark critical thinking.

 ### Inspire Investigation
Spark inquiry-driven, hands-on exploration.

100% Built for Next Generation Science Standards

 ### Inspire Innovation
Spark creative solutions to real-world challenges.

Get started with your Program Guide!

This user-friendly guide will help you get familiar with the program philosophy and design, module and lesson structure, and digital experience.

Need login credentials?

Go to https://my.mheducation.com and select "Create an account."

Program Authors

Dr. Jay Hackett

Dr. Jay Hackett is an emeritus professor of Earth Sciences and past recipient of the William R. Ross Science Award as an Honored Alumnus at the University of Northern Colorado. Dr. Hackett is co-author of *Teaching Science as Investigations* and made contributions to the development of *Inquiry and the National Science Education Standards: A Guide for Teaching and Learning*. Dr. Hackett is an admired science educator and McGraw-Hill Education science author.

Page Keeley, M.Ed.

Page Keeley, M.Ed. is a nationally-renowned expert on science formative assessment and teaching for conceptual change. She is the author of several award-winning books and journal articles on uncovering student thinking using formative assessment probes and techniques. She was the Science Program Director at the Maine Mathematics and Science Alliance for 16 years and a past President of the National Science Teachers Association. Currently she is an independent consultant providing professional development to school districts and science education organizations and a frequent invited speaker at national conferences.

Dr. Jo Anne Vasquez

Dr. Jo Anne Vasquez, a past President of the National Science Teachers Association and the National Science Education Leadership Association, was the first elementary educator to become a Presidential Appointee to the National Science Board, the governing board of the National Science Foundation. Her distinguished service and extraordinary contributions to the advancement of science and STEM education at the local, state, and national levels has won her numerous awards: 2014 National Science Education Leadership Award for Outstanding Leadership in Science Education, 2013 National Science Board Public Service Award, and "Robert H. Carlton Award" for Leadership in Science Education.

Dr. Richard Moyer

Dr. Richard Moyer is an emeritus professor of Science Education and Natural Sciences at the University of Michigan-Dearborn. He is an award-winning educator, author, and co-author of *Everyday Engineering: Putting the E in STEM Teaching and Learning, Teaching Science as Investigation,* and *More Everyday Engineering*. Dr. Moyer has served for over 33 years as a McGraw-Hill Education science author.

Dr. Dorothy J.T. Terman

Dr. Dorothy J.T. Terman served for 21 years as Science Coordinator for California's Irvine Unified School District, responsible for science curriculum development, program implementation, and assessment. She holds a B.S. in Science Education from Cornell University, an M.A. in Cell Biology from Columbia University, and a Ph.D. in Curriculum from the University of Iowa. She has received many awards, including the Ohaus Award from the National Science Teachers Association for Innovation in Elementary Science Education. She is a consultant for inquiry-based science curriculum implementation and a veteran McGraw-Hill Education science author.

Dinah Zike, M.Ed.

Dinah Zike, M.Ed. is an award-winning author, educator, and inventor known for designing three-dimensional hands-on manipulatives and graphic organizers known as Foldables® and VKVs® (Visual Kinesthetic Vocabulary®). Ms. Zike is the founder and President of Dinah-Might Adventures, LP and Dinah Zike Academy. She is also the recipient of the Teachers' Choice Award For the Classroom and Teachers' Choice Award For Professional Development.

Learning Science Research Council

Our Learning Science Research Council (LSRC) was founded to promote, collaborate, and seek funding for research that will result in new and improved products with significant impact on student learning outcomes. The LSRC is comprised of McGraw-Hill Education professionals whose areas of expertise are built upon learning science research and are informed by the productive collaborations with researchers inside and outside MHE. The LSRC has invited a number of highly regarded scholars from the United States and abroad to be members of an Advisory Board to the LSRC. Members of the Advisory Board are renowned learning scientists who can contribute, either directly or through their networks and convening power, to the LSRC's agenda and vision.

Hero Images/Getty Images

Key Partners

The Concord Consortium is a nonprofit educational research and digital learning organization focused on delivering the promise of technology for education in science, math, and engineering. The *Inspire Science* simulations, created in partnership with The Concord Consortium, enable students to model concepts otherwise not possible to explore in the classroom.

Filament Games creates digital learning games and interactives designed to foster 21st-century skills through experiential learning. The immersive games included with *Inspire Science*, developed in partnership with Filament Games, enable students to "play" with the lesson concepts to deepen conceptual understanding.

Program Advisors

Phil Lafontaine
NGSS Educational Consultant
Sacramento, California

Emily C. Miller
University of Wisconsin at Madison
Madison, Wisconsin

Dr. Timothy Shanahan
Distinguished Professor Emeritus
University of Illinois at Chicago
Chicago, Illinois

Jody Skidmore Sherriff
Regional Director
K-12 Alliance/WestEd
Sacramento, California

Content Reviewers

Dr. Cindy Klevickis
Professor of Integrated Science
and Technology
James Madison University
Harrisonburg, Virginia

Derrick Svelnys, M.S., M.Ed.
Teacher
Chicago Public Schools
Chicago, Illinois

Kandi K. Wojtysiak, M.Ed.
Science Department Chair
Notre Dame Preparatory High School
Scottsdale, Arizona

Teacher Reviewers

Jennifer Covarrubias
Kinetic Academy
Huntington Beach, California

Amy Syverson Kunis, M.A.
Perris Elementary School District
Perris, California

Tasha Terrill, M.Ed.
Highland Local School District
Sparta, Ohio

Monica Galavan, M.A., M.S.
Cajon Valley Union School District
San Diego, California

Teresa Harris-Belcher, B.A.
Highland Local School District
Sparta, Ohio

Kathi Lundstrom, Ph.D.
Norwalk-La Mirada Unified School District
La Mirada, California

Amanda Waggoner
Highland Local School District
Sparta, Ohio

Mika L. George, B.S.
Highland Local School District
Sparta, Ohio

Lisa K. Reely
Highland Local School District
Sparta, Ohio

Kimberly Wilson, M.Ed.
Mariners Christian School
Costa Mesa, California

Table of Contents

Module: Plant Parents and Their Offspring

Lesson 1: Plants and Their Parents

Lesson 2: Plant Survival

Table of Contents

Teacher Notes

Notes

Performance Expectations at a Glance

In this unit, students will discover and practice the Science and Engineering Practices, Disciplinary Core Ideas, and Crosscutting Concepts needed to perform the following Performance Expectations.

Performance Expectations	MODULE: Plant Structures and Functions	MODULE: Plant Parents and Their Offspring
K-2-ETS1-2	●	
K-2-ETS1-3		●
1-ESS1-1	●	
1-LS1-1	●	●
1-LS3-1		●
1-PS4-3	●	

 ## Correlations by Module to the NGSS

MODULE: Plant Structures and Functions

K-2-ETS	Engineering Design	
K-2-ETS1-2	Develop a simple sketch, drawing, or physical model to illustrate how the shape of an object helps it function as needed to solve a given problem.	42, *43–44*

SEP Science and Engineering Practices

Developing and Using Models Modeling in K–2 builds on prior experiences and progresses to include using and developing models (i.e., diagram, drawing, physical replica, diorama, dramatization, or storyboard) that represent concrete events or design solutions. • Develop a simple model based on evidence to represent a proposed object or tool. (K-2-ETS1-2)	*38–39*, 42, *43–44*

DCI Disciplinary Core Ideas

ETS1.B: Developing Possible Solutions • Designs can be conveyed through sketches, drawings, or physical models. These representations are useful in communicating ideas for a problem's solutions to other people. (K-2-ETS1-2)	42, *43–44*

Inquiry activities are in italics.

Continued from previous page.

CCC Crosscutting Concepts	
Structure and Function • The shape and stability of structures of natural and designed objects are related to their function(s). (K-2-ETS1-2)	2–3, *10–12*, 20–21, 25, 32–33, 34–35, 38–39, 42, *43–44*

1-LS1	**From Molecules to Organisms: Structures and Processes**	
1-LS1-1	**Use materials to design a solution to a human problem by mimicking how plants and/or animals use their external parts to help them survive, grow, and meet their needs.*** *[Clarification Statement: Examples of human problems that can be solved by mimicking plant or animal solutions could include designing clothing or equipment to protect bicyclists by mimicking turtle shells, acorn shells, and animal scales; stabilizing structures by mimicking animal tails and roots on plants; keeping out intruders by mimicking thorns on branches and animal quills; and, detecting intruders by mimicking eyes and ears.]*	42, *43–44*

SEP Science and Engineering Practices	
Constructing Explanations and Designing Solutions Constructing explanations and designing solutions in K–2 builds on prior experiences and progresses to the use of evidence and ideas in constructing evidence-based accounts of natural phenomena and designing solutions. • Use materials to design a device that solves a specific problem or a solution to a specific problem. (1-LS1-1)	*38–39*, 41, 42, *43–44* Teacher's Edition *Only*: 29

DCI Disciplinary Core Ideas	
LS1.A: Structure and Function • All organisms have external parts. Different animals use their body parts in different ways to see, hear, grasp objects, protect themselves, move from place to place, and seek, find, and take in food, water and air. Plants also have different parts (roots, stems, leaves, flowers, fruits) that help them survive and grow. (1-LS1-1)	2–3, 7, *10–12*, 13, 14–15, 16–17, *18–19*, 20–21, 23, 25, 32–33, 34–35, *38–39*, 41, 45 Teacher's Edition *Only*: 8, 26, 40
LS1.D: Information Processing • Animals have body parts that capture and convey different kinds of information needed for growth and survival. Animals respond to these inputs with behaviors that help them survive. Plants also respond to some external inputs. (1-LS1-1)	26–27, *28–31*, 41

Inquiry activities are in italics.

Continued from previous page.

CCC Crosscutting Concepts

Structure and Function • The shape and stability of structures of natural and designed objects are related to their function(s). (1-LS1-1)	2–3, 7, *10–12*, 13, 14–15, 16–17, *18–19*, 20–21, 23, 25, 32–33, 34–35, *38–39*, 41, 42
Connections to Engineering, Technology, and Applications of Science **Influence of Science, Engineering and Technology on Society and the Natural World.** • Every human-made product is designed by applying some knowledge of the natural world and is build using materials derived from the natural world. (1-LS1-1)	36–37, *43–44* Teacher's Edition *Only*: 35

CCSS Math Connections

1.MD.A.2	*19*

ELD Connections

ELD.PII.1.6	19, 21, 22, 33, 39, 40

CCSS ELA/Literacy Connections

W.1.8	42

ALSO INTEGRATES:

1-ESS1-1	26–27, *28–31*
1-PS4-3	*10–12* Teacher's Edition *Only*: 11
SEP Analyzing and Interpreting Data	*18–19, 28–31, 38–39* Teacher's Edition *Only*: 11
SEP Planning and Carrying Out Investigations	*38–39*
DCI ESS1.A	26–27, *28–31*
DCI PS4.B	*10–12*
CCC Patterns	Teacher's Edition *Only*: 8, 15, 23, 29
ELA RL.1.5	15
ELA RL1.7	*12*

Inquiry activities are in italics.

MODULE: Plant Parents and Their Offspring

K-2-ETS	Engineering Design	
⬤ K-2-ETS1-3	Analyze data from tests of two objects designed to solve the same problem to compare the strengths and weaknesses of how each performs.	*87–88*

SEP Science and Engineering Practices

Analyzing and Interpreting Data Analyzing data in K–2 builds on prior experiences and progresses to collecting, recording, and sharing observations. • Analyze data from tests of an object or tool to determine if it works as intended. (K-2-ETS1-3)	*54–57, 62–63, 87–88*

DCI Disciplinary Core Ideas

ETS1.C: Optimizing the Design Solution • Because there is always more than one possible solution to a problem, it is useful to compare and test designs. (K-2-ETS-1-3)	*87–88*

1-LS1	From Molecules to Organisms: Structures and Processes	
⬤ 1-LS1-1	Use materials to design a solution to a human problem by mimicking how plants and/or animals use their external parts to help them survive, grow, and meet their needs.* [Clarification Statement: Examples of human problems that can be solved by mimicking plant or animal solutions could include designing clothing or equipment to protect bicyclists by mimicking turtle shells, acorn shells, and animal scales; stabilizing structures by mimicking animal tails and roots on plants; keeping out intruders by mimicking thorns on branches and animal quills; and, detecting intruders by mimicking eyes and ears.]	80, *82–83*, *87–88* Teacher's Edition *Only*: 76

SEP Science and Engineering Practices

Constructing Explanations and Designing Solutions Constructing explanations and designing solutions in K–2 builds on prior experiences and progresses to the use of evidence and ideas in constructing evidence-based accounts of natural phenomena and designing solutions. • Use materials to design a device that solves a specific problem or a solution to a specific problem. (1-LS1-1)	80, *82–83*, 85, *87–88* Teacher's Edition *Only*: 73

Inquiry activities are in italics.

Continued from previous page.

DCI Disciplinary Core Ideas	
LS1.A: Structure and Function • All organisms have external parts. Different animals use their body parts in different ways to see, hear, grasp objects, protect themselves, move from place to place, and seek, find, and take in food, water and air. Plants also have different parts (roots, stems, leaves, flowers, fruits) that help them survive and grow. (1-LS1-1)	*54–57*, 75, 78–79, 80, *82–83*, 85 Teacher's Edition *Only*: 46, 70
LS1.D: Information Processing • Animals have body parts that capture and convey different kinds of information needed for growth and survival. Animals respond to these inputs with behaviors that help them survive. Plants also respond to some external inputs. (1-LS1-1)	*72–73*, 80 Teacher's Edition *Only*: 46, 69

CCC Crosscutting Concepts	
Structure and Function • The shape and stability of structures of natural and designed objects are related to their function(s). (1-LS1-1)	76–77, 78–79, 80, *82–83*, 85 Teacher's Edition *Only*: 74
Connections to Engineering, Technology, and Applications of Science **Influence of Science, Engineering and Technology on Society and the Natural World.** • Every human-made product is designed by applying some knowledge of the natural world and is build using materials derived from the natural world. (1-LS1-1)	80, *87–88*

1-LS3	**Heredity: Inheritance and Variation of Traits**	
1-LS3-1	**Make observations to construct an evidence-based account that young plants and animals are like, but not exactly like, their parents.** *[Clarification Statement: Examples of patterns could include features plants or animals share. Examples of observations could include leaves from the same kind of plant are the same shape but can differ in size; and, a particular breed of dog looks like its parents but is not exactly the same.] [Assessment Boundary: Assessment does not include inheritance or animals that undergo metamorphosis or hybrids.]*	*54–57*, 58–59, 60–61

SEP Science and Engineering Practices	
Constructing Explanations and Designing Solutions Constructing explanations and designing solutions in K–2 builds on prior experiences and progresses to the use of evidence and ideas in constructing evidence-based accounts of natural phenomena and designing solutions. • Make observations (firsthand or from media) to construct an evidence-based account for natural phenomena. (1-LS3-1)	*54–57*, 62–63, 67, *82–83*

Inquiry activities are in italics.

Continued from previous page.

DCI Disciplinary Core Ideas	
LS3.A: Inheritance of Traits • Young animals are very much, but not exactly like, their parents. Plants also are very much, but not exactly, like their parents. (1-LS3-1)	51, 52–53, *54–57*, 58–59, 60–61, *62–63*, 67
LS3.B: Variation of Traits • Individuals of the same kind of plant or animal are recognizable as similar but can also vary in many ways. (1-LS3-1)	52–53, *54–57*, 60–61, *62–63*, 67 Teacher's Edition *Only*: 58
CCC Crosscutting Concepts	
Patterns • Patterns in the natural and human designed world can be observed, used to describe phenomena, and used as evidence. (1-LS3-1)	*62–63*, 67 Teacher's Edition *Only*: 55
CCSS Math Connections	
1.MD.A.2	65, 87
ELD Connections	
ELD.PII.1.6	61
CCSS ELA/Literacy Connections	
W.1.8	57, 82
ALSO INTEGRATES:	
K-2-ETS1-1	80
SEP Developing and Using Models	79
SEP Planning and Carrying Out Investigations	68
ELD.LS1.A	82
ELD.PI.1.1	57, 77
RI.1.3	77

Inquiry activities are in italics.

Three-Dimensional Learning

In this module, students will investigate the functions of plant structures and design a solar-powered light stand.

SEP Science and Engineering Practices

- Constructing Explanations and Designing Solutions
- Developing and Using Models

DCI Disciplinary Core Ideas

- ETS1.B Developing Possible Solutions
- LS1.A Structure and Function
- LS1.D Information Processing

CCC Crosscutting Concepts

- *Connections to Engineering, Technology, and Applications of Science*
  Influence of Science, Engineering and Technology on Society and the Natural World
- Structure and Function

Performance Expectations

1-LS1-1. Use materials to design a solution to a human problem by mimicking how plants and/or animals use their external parts to help them survive, grow, and meet their needs.

1-ESS1-1. Use observations of the sun, moon, and stars to describe patterns that can be predicted.*

1-PS4-3. Plan and conduct investigations to determine the effect of placing objects made with different materials in the path of a beam of light.*

K-2-ETS1-2. Develop a simple sketch, drawing, or physical model to illustrate how the shape of an object helps it function as needed to solve a given problem.

*Performance expectations that are introduced but not assessed until later units.

CROSS-CURRICULAR ▶ Connections

In addition to in-depth coverage of the three dimensions, this module also covers connections to Math, English-Language Arts, and Environmental topics.

⊘ **GO ONLINE** for Professional Learning videos that support three-dimensional learning.

Disciplinary Core Idea Progressions

K-2	3-5	6-8
ETS1.B Developing Possible Solutions		
• Designs can be conveyed through sketches, drawings, or physical models. These representations are useful in communicating ideas for a problem's solutions to other people. (K-2-ETS1-2)	• Research on a problem should be carried out before beginning to design a solution. Testing a solution involves investigating how well it performs under a range of likely conditions. (3-5-ETS1-2) • At whatever stage, communicating with peers about proposed solutions is an important part of the design process, and shared ideas can lead to improved designs. (3-5-ETS1-2)	• A solution needs to be tested, and then modified on the basis of the test results, in order to improve it. (MS-ETS1-4) • There are systematic processes for evaluating solutions with respect to how well they meet the criteria and constraints of a problem. (MS-ETS1-2), (MS-ETS1-3) • Sometimes parts of different solutions can be combined to create a solution that is better than any of its predecessors. (MS-ETS1-3) • Models of all kinds are important for testing solutions. (MS-ETS1-4)
LS1.A Structure and Function		
• All organisms have external parts. Different animals use their body parts in different ways to see, hear, grasp objects, protect themselves, move from place to place, and seek, find, and take in food, water and air. Plants also have different parts (roots, stems, leaves, flowers, fruits) that help them survive and grow. (1-LS1-1)	• Plants and animals have both internal and external structures that serve various functions in growth, survival, behavior, and reproduction. (4-LS1-1)	• All living things are made up of cells, which is the smallest unit that can be said to be alive. An organism may consist of one single cell (unicellular) or many different numbers and types of cells (multicellular). (MS-LS1-1) • Within cells, special structures are responsible for particular functions, and the cell membrane forms the boundary that controls what enters and leaves the cell. (MS-LS1-2) • In multicellular organisms, the body is a system of multiple interacting subsystems. These subsystems are groups of cells that work together to form tissues and organs that are specialized for particular body functions. (MS-LS1-3)
LS1.D Information Processing		
• Animals have body parts that capture and convey different kinds of information needed for growth and survival. Animals respond to these inputs with behaviors that help them survive. Plants also respond to some external inputs. (1-LS1-1)	• Different sense receptors are specialized for particular kinds of information, which may be then processed by the animal's brain. Animals are able to use their perceptions and memories to guide their actions. (4-LS1-2)	• Each sense receptor responds to different inputs (electromagnetic, mechanical, chemical), transmitting them as signals that travel along nerve cells to the brain. The signals are then processed in the brain, resulting in immediate behaviors or memories. (MS-LS1-8)

Three Dimensions at a Glance

Throughout this module and in the culminating module project, students will integrate relevant Science and Engineering Practices and Crosscutting Concepts into their learning and understanding of the Disciplinary Core Ideas. Use this chart to locate where students will encounter each of the three dimensions that build to the Performance Expectations.

DIMENSIONS	LESSON 1	LESSON 2	MODULE PROJECT
SEP Constructing Explanations and Designing Solutions (1-LS1-1)	•	•	
SEP Developing and Using Models (K-2-ETS1-2)			•
DCI ETS1.B Developing Possible Solutions (K-2-ETS1-2)			•
DCI LS1.A Structure and Function (1-LS1-1)	•	•	•
DCI LS1.D Information Processing (1-LS1-1)		•	
CCC *Connections to Engineering, Technology, and Applications of Science* Influence of Science, Engineering and Technology on Society and the Natural World (1-LS1-1)		•	
CCC Structure and Function (1-LS1-1, K-2-ETS1-2)	•	•	•

Module Planner

In this module, students learn the function of plant structures.

	Module Opener	Lesson 1: Plant Parts	Lesson 2: Functions of Plant Parts
	Big Idea: What are the functions of common plant structures?	**Essential Question:** What patterns can you find between different plants?	**Essential Question:** What do plant structures do?
Pacing 1 Day = 45 min	0.5 Day	7 Days	6 Days
Summary	In this module, students will learn how plant structures help plants live.	Students will learn characteristics of common plant structures.	Students will learn the function of plant structures.
Inquiry Activity		**Hands On** Observe Plant Parts **Hands On** Plant Structures	**Hands On** Plants and Light **Hands On** Celery Stems
Vocabulary		structure	flower, fruit, function, leaf, root, seed, stem
Cross-Curricular Connections		ELA, Math	ELA

School-to-Home Resources

🔘 **GO ONLINE** for the following resources to strengthen the school-to-home connections.

Letter to Home will help parents and guardians understand the learning objectives for the Plant Structures and Functions module.

STEM Module Project: Build a Solar-Powered Light Stand	Module Wrap-Up
2 Days	0.5 Day
Students will use what they've learned throughout this module to design and build a model of a solar-powered light stand. They will explain how their light stand is similar to a plant structure and function.	Students will revisit the module phenomenon and explain their learning.
Engineering Connection: Build a Solar-Powered Light Stand	

Assessment Tools

Formative Assessment
Includes Page Keeley Science Probes; Claim, Evidence, Reasoning; three dimensional learning checks

STEM Module Project
Authentic performance-based assessment with rubric

McGraw-Hill Assessment
Ready-made assessments that can be printed or delivered electronically

Module Planner

In this module, students will learn about the function of common plant structures. They will use this information to develop a model of a solar-powered light stand.

Lesson	Inquiry Activity		Materials	
	★ 🔵 **GO ONLINE** for teacher support videos on selected activities.		**Consumable**	**Non-Consumable**
	Materials included in the Collaboration Kit are listed in blue.			
Lesson 1	**Hand On** Observe Plant Parts	🕐 30 min ⛹ small group	plant, crayons, plastic gloves, batteries	hand lens, flashlight
	Purpose: Students will observe that plants are made of many separate parts.			
	Hands On Plant Structures	🕐 30 min ⛹ small group	onion, daisy, plastic gloves	hand lens, cubes
	Purpose: Students will compare an onion and a daisy to find which structures they have in common.			
Lesson 2	**Hands On** Plants and Light	🕐 30 min ⛹ small group	plant, crayons, plastic gloves	
	Purpose: Students will observe how a plant moves in response to the Sun.			
	★ **Hands On** Celery Stems	🕐 30 min ⛹ small group	blue food coloring, water, celery stalk, plastic gloves	clear container, hand lens
	Purpose: Students will observe how water moves through celery to verify the function of a stem.			
Module Project	★ **Engineering Challenge** Build a Solar-Powered Light Stand	🕐 60 min ⛹ small group	Suggested: tissue box, paper, glue, cardboard tubes, chenille stems, tape	Suggested: scissors
	Purpose: Students will use the engineering design process and what they have learned about plants to design and build a solar-powered light stand.			

McGraw-Hill Education is your partner for hands-on materials! To order new Collaboration Kits or refill specific items, contact the McGraw-Hill Education customer support line at (800) 338-3987.

Inquiry Activity Support

Guide activities with confidence by watching the Inquiry Activity Teacher Preview video as you plan the day's Inquiry Activity. After your students complete the activity, give them all a common set of expected observations by showing them the Inquiry Rewind video.

Inquiry Activity Teacher Preview

INQUIRY ACTIVITY

As you plan each Explore Inquiry Activity, watch this video for information about activity setup, strategies for a smooth activity experience, and math and skills reviews you can do to launch the activity and prepare your students to complete the activity successfully.

Inquiry Activity Rewind

Each Inquiry Activity Rewind video shows students a step-by-step procedure and expected or sample results. Highlight important observations students might have missed during the activity so, even if they missed class, everyone is ready to interpret the data and construct explanations.

Teacher Notes

Inspire All Students

Strategies to scaffold your instruction and plan for successful teaching for all students.

Differentiated Instruction

Plants use their structures to survive, grow and meet their needs. People can mimic the structures of plants to solve their own problems and needs. Help students to connect these concepts by providing multiple means of expression.

AL Approaching Level

As students learn new information about plant parts, have them analyze some pictures of man-made objects that have similar structures and function. EX: Show a picture of a flag pole Ask: What problem does the pole solve? How is this like part of a plant?

OL On Level

Use scrap and recycled materials (EX: straws, paper towel rolls, pipe cleaners, foil, construction paper, string) have students create models of plant parts. Ask them to tell how the materials they chose relate to the structure and function of plant parts.

BL Beyond Level

As students learn new information about plant parts, have them go online to find pictures of man-made objects with similar structure and function to plant parts. Have them share and explain the connections they made with the class.

Advanced Learners and Gifted Learners

Instruction should focus on adding depth and complexity in asking questions about the structure and function of plants and planning ways to investigate them.

DOK 3 Strategic Thinking Have students revisit the investigations they conduct throughout the module and have them ask further questions. EX: What would happen if we moved the plants further away from the window? Have them write or draw their plans for further investigation.

DOK 4 Extended Thinking Provide students with plant-related materials (artificial plants, books, play soil, etc.). Have them conduct their own investigations (EX: What will happen if I put this flower in play soil? What flower parts will I see in these books?) Have them tell about their process.

Literacy Support: Using the Leveled Readers

Use the Leveled Readers to enable students to further develop their literacy skills through science.

- Fiction: Engages students in key concepts
- Nonfiction: Focuses on real-world topics; Makes informational text accessible to all learners
- Also available in print and online.

Parts of Plants

Summary This book describes the function of common plant structures.

When to Use Use during Lesson 1 to further develop an understanding of plant structures.

Hayri Er/E+/Getty Images

English-Language Support

Utilize charts and graphic organizers to help students collect and use information that will help them design solutions that mimic the parts of a plant. Throughout the module, guide them to notice solutions to problems such as: Staying upright, staying in the ground, getting enough sunlight.

EMERGING

Concept Map Create a chart of plant needs with the students. Write a different plant need or problem in each column. Have students draw about different plant parts on sticky notes as they come across them in the module. Have them match the plant parts to the problems they solve.

EXPANDING

Word Wall Add words about important plant parts to a word wall as you come across them in the module. Ask students to write about how the plant parts help solve a plant's problem using sentence prompts such as: ____ Sample answer: leaves help a plant ____ Sample answer: make food from sunlight.

BRIDGING

T-Chart Brainstorm possible problems or needs that plants have. Have students list these on the left side of the T-chart. Have them collect information about how those needs are met on the right side of the chart

Cognates

Cognates are words in two different languages that share a similar meaning, spelling, and pronunciation. Review differences in spelling and pronunciation of these terms with your Spanish-speaking English learners.

fruit	survive	mineral
fruta	sobrevivir	mineral

Vocabulary Resources

The online Vocabulary Resources are intended to support English language learning and vocabulary acquisition. Resources for each stage of the learning process are meant to appeal to different types of learners.

Explore	Study	Review
Vocabulary Concept Circle	Frayer Model	Science Vocabulary

Performance Expectations

The learning experiences throughout this module will develop student understanding of the following Performance Expectations:

1-ESS1-1 Use observations of the sun, moon, and stars **to describe patterns** that can be predicted.

1-LS1-1 Use materials to design a solution to a human problem by mimicking how plants and/or animals use their external parts to survive, grow, and meet their needs.

1-PS4-3 Plan and conduct investigations to determine the effect of placing objects made with different materials in the path of a beam of light.

K-2-ETS1-2 Develop a simple sketch, drawing, or physical model to illustrate how the shape of an object helps it function as needed to solve a given problem.

Plant Structures and Functions

2   **Module:** Plant Structures and Functions

Teacher Toolbox

Science Background

This photo shows a young tree. The above and below ground structures are clearly visible. This perspective provides a unique look at how a plant is anchored into the ground by its roots.

Identifying Preconceptions

Students may not realize that in addition to the structures seen above ground, plants also have structures below ground. Without structures like stems and roots, plants would not be able to collect the water and nutrients they need.

A common preconception is that plants do not have large structures underground. While this may be the case for plants like moss, other plants like prairie grass have elaborate underground networks of roots that are larger than the plant above ground. As students explore this lesson, emphasize that plants come in thousands of varieties, but that patterns exists among plants and plant structures.

DISCOVER
THE PHENOMENON

How does this plant stay upright?

 Plant Parts

 GO ONLINE

Check out *Plant Parts* to see this phenomenon in action.

 Talk About It

Look at the photo.
Watch the video.
What do you observe?
What questions do you have?

Did You Know?

In some plants, the parts underground are larger than the parts we can see above the soil!

Module: Plant Structures and Functions **3**

 GO ONLINE

ThomasVogel/E+/Getty Images

DISCOVER THE PHENOMENON

🕐 10 min   👥 whole class

Science often begins when someone makes an observation about a situation or occurrence. Scientists refer to an event or situation that is observed or can be studied as a phenomenon. Have students study the picture of the young tree with its above and below ground structures visible.

Ask the **Discover the Phenomenon** question:

How does this plant stay upright?

➡️ This leads to the overarching module **Big Idea:**

What are the functions of common plant structures?

 GO ONLINE Check out *Plant Parts* to see the phenomenon in action.

💬 Talk About It

Ask students to describe what they see. Help students turn their observations from the video into questions. Record observations on the board or chart paper. Start a class discussion with the following prompts:

• Why do you think plants have parts below the ground?

• What parts of a plant do you know?

Record responses and questions on the board or chart paper to look at as you move through this module.

Did You Know?

Plants such as buffalo grass and potatoes often have underground structures larger or heavier than their aboveground structures.

Differentiated Instruction

AL Guide the students to ask questions and define problems by creating a KWL chart. Revisit the chart throughout the lesson to write what students "Learned."

OL Activate students' background knowledge of problems plants face by comparing areas where they have seen plants grow to areas where they have not.

BL Have students share their thinking about how the plant stays upright. Have them brainstorm ways they can investigate how a plant solves this problem.

What Does a Landscape Architect Do?

 10 min   whole class

Introduce the landscape architect STEM CAREER Connection Encourage students to look at the photos and share their observations. Have students read question 1 and discuss their answers with the class.

Throughout the module and lessons, students should apply the skills of a landscape architect to their learning. Remind students that landscape architects use their understanding of plant structures and functions in their designs.

GO ONLINE to watch the video *Landscape Architects* to learn more about this STEM career.

STEM CAREER Connection

What Does a Landscape Architect Do?

GO ONLINE
Learn about a
Landscape Architect.

Landscape architects design outdoor spaces. Parks and playgrounds are outdoor spaces. Landscape architects must know how different plants grow.

1. What do landscape architects need to know about plants?

Sample answer: Landscape architects need to know how much sunlight and water plants need. They need to know which plants grow the best in California.

4   STEM CAREER Connection   **Module:** Plant Structures and Functions

---

GO ONLINE

INTERACTIVE PRESENTATION

STEM Career: Landscape Architect

STEM CAREER

2. Think about an outdoor space where you live. Draw a picture of it. Name the outdoor space.

> Drawing should include an outdoor space and a title. Drawing could include plants, outdoor furniture, play equipment, or a body of water.

ENVIRONMENTAL Connection

3. Explain something a landscape architect can do to make sure outdoor spaces stay safe and healthy.

A landscape architect can put garbage cans in the spaces to keep them clean.

Tell a partner your ideas.

KAYLA
Landscape Architect

STEM CAREER Connection   **Module:** Plant Structures and Functions   **5**

Have students answer question 2 and share their drawings with classmates. Encourage students to explain the reasons behind their design. Ask students why they decided to put plants in specific places in their outdoor space.

ASK: What living things are in the outdoor space you drew? Sample answer: There are flowers, grass, and people in my outdoor space.

ASK: What nonliving things are in the outdoor space you drew? Sample answer: There are benches, sidewalks, and a pond in my outdoor space.

ASK: How do the plants in your outdoor space help the animals and people? Sample answer: The tree in my outdoor space gives shade to the people and a home for animals like squirrels and birds.

Begin a discussion about the ways people can change outdoor spaces like parks, forests, and beaches.

ASK: How do people help make outdoor spaces better? Sample answer: People can help outdoor spaces by keeping pollution out of the water and soil. They can also help protect plants and animals.

ASK: How do people make outdoor spaces worse? Sample answer: Sometimes people leave garbage in outdoor spaces. This can hurt the plants and animals that live there.

Read Kayla's speech bubble to students. As a class, discuss how landscape architects can help keep outdoor spaces safe and healthy.

Word Wall

 10 min whole class

The words shown on the word wall represent some of the fundamental words from this module. Additional vocabulary words are introduced in the module lessons. By the end of this module, students should be able to use these module vocabulary words correctly in context.

Write each vocabulary word on the board. As you write each word, have students pronounce it. Have students describe the photo next to each word to get them thinking about what the word might mean. Record student responses on the board next to the word. Revisit the terms throughout the module as students learn each meaning in context.

STEM Vocabulary

As students encounter each word, have them use context clues to determine the meaning. Provide the following background information to help students develop a clear understanding of these terms. Help students build their academic vocabulary by recording these words on a board.

investigate In first grade, students are expected to plan and carry out investigations. To build familiarity with this term, encourage students to investigate answers to questions as they arise in class.

ASK: How can we investigate what temperature it is outside? Sample answer: We can take a walk outside and feel if it is hot or cold. We can use a thermometer to take the temperature. We can also look up the weather on a computer.

observe In first grade, students should be able to make, record, and share observations. Ask students to look around the room and describe an object they observe. Ask students to use other senses to add detail to their descriptions. This can be turned into a game. Pick an object in the room to observe. Make observations about the object and share them with the class. Do not tell students what object you are looking at. Have students guess what object you are describing. Have students take turns observing and guessing with partners.

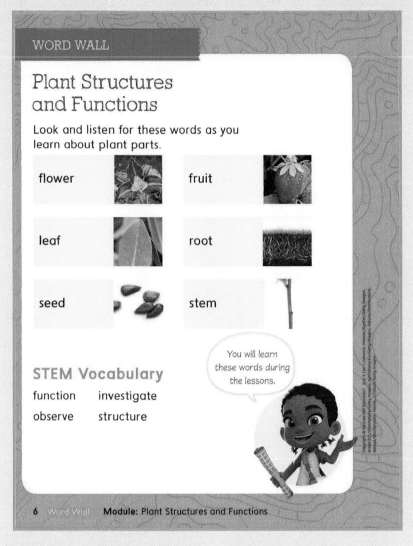

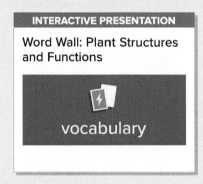 GO ONLINE

INTERACTIVE PRESENTATION

Word Wall: Plant Structures and Functions

vocabulary

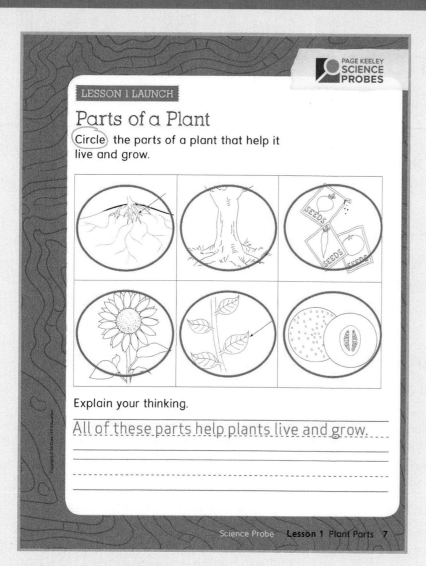

Parts of a Plant

(Circle) the parts of a plant that help it live and grow.

Explain your thinking.

All of these parts help plants live and grow.

🡒 GO ONLINE

INTERACTIVE PRESENTATION

Science Probe: Parts of a Plant

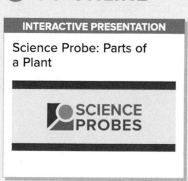

PAGE KEELEY
SCIENCE PROBES

Parts of a Plant

🕒 10 min   👥 whole class

Using the Probe

The purpose of this probe is to identify students' prior knowledge about parts of plants. This probe works well with a think-pair-share strategy. Have students think about and mark their answers in their notebook. Next, have students share their ideas with a partner. Lastly, call on a few students to share their ideas with the class. Give students time to make changes to their original ideas when they record their explanations. Use this probe to assess students' prior knowledge of the lesson content and to identify possible misconceptions.

Be sure not to tell students the answer. It is not important that students know the answer to this probe at this point in the lesson. What is important is the reasoning students provide to support their answer. Students will revisit the probe throughout the lesson to see how their thinking has changed.

Throughout the Lesson

Use students' explanations to bridge the students' initial ideas about plant parts with the understanding they will develop. Prompts in the Teacher's Edition will instruct you when it's time for students to revisit the probes.

Teacher Explanation

Students should have circled all of the pictures. Roots, trunks, seeds, flowers, leaves, and fruit are all commom plant structures. Each of the parts serve a different function in helping the plant carry out its life functions.

Teacher Toolbox

Identifying Preconceptions

Students may think that plants are simple organisms when compared with animals when, in fact, they are complex. Students may not realize that a tree is a plant. Therefore they may not understand that a trunk is a plant part. Likewise, students may not realize that the fruits they eat are also parts of plants.

Building to the Performance Expectations:

In this lesson, students will explore content and develop skills leading to mastery of the following Performance Expectation(s):

1-LS1-1. Use materials to design a solution to a human problem by mimicking how plants and/or animals use their external parts to help them survive, grow, and meet their needs.

1-PS4-3. Plan and conduct investigations to determine the effect of placing objects made with different materials in the path of a beam of light.*

*This Performance Expectation is introduced but not assessed in its entirety during this lesson.

SEP Science and Engineering Practices

Constructing Explanations and Designing Solutions

Constructing explanations and designing solutions in K–2 builds on prior experiences and progresses to the use of evidence and ideas in constructing evidence-based accounts of natural phenomena and designing solutions. (1-LS1-1)

DCI Disciplinary Core Idea

LS1.A Structure and Function

All organisms have external parts. Different animals use their body parts in different ways to see, hear, grasp objects, protect themselves, move from place to place, and seek, find, and take in food, water and air. Plants also have different parts (roots, stems, leaves, flowers, fruits) that help them survive and grow. (1-LS1-1)

LS1.D Information Processing

Animals have body parts that capture and convey different kinds of information needed for growth and survival. Animals respond to these inputs with behaviors that help them survive. Plants also respond to some external inputs. (1-LS1-1)

CCC Crosscutting Concept

Structure and Function

The shape and stability of structures of natural and designed objects are related to their function(s). (1-LS1-1)

***Connections to Engineering, Technology, and Applications of Science** Influence of Science, Engineering and Technology on Society and the Natural World*

Every human-made product is designed by applying some knowledge of the natural world and is built using materials derived from the natural world. (1-LS1-1)

Reading Connections	Math Connections
RL.1.5, RL.1.7	1.MD.A.2 Measurement and Data

* See correlation table for full text of ELA and Math standards.

Track Your Progress to the Performance Expectations

You may want to return after completing the lesson to note concepts that will need additional review before your students start the module Performance Project.

Dimension	Concepts to Review Before Assessment
SEP Constructing Explanations and Designing Solutions (1-LS1-1)	
DCI LS1.A Structure and Function (1-LS1-1)	
DCI LS1.D Information Processing (1-LS1-1)	
CCC Connections to Engineering, Technology, and Applications of Science Influence of Science, Engineering of Technology on Society and the Natural World (1-LS1-1)	
CCC Structure and Function (1-LS1-1)	

Lesson at a Glance

Full Track is the recommended path for the complete lesson experience. FlexTrack A and FlexTrack B provide timesaving strategies and alternatives.

	Day-to-Day	Pacing	Resources
		Full Track 45 min/day (full year)	
Assess Prior Knowledge	Page Keeley Science Probes: *Parts of a Plant*	Day 1	Page 7
Engage	Discover the Phenomenon: How is this sequoia tree different from other plants?		Pages 8–9 Video: *Sequoia National Park*
Explore	Inquiry Activity: *Observe Plant Parts*	Day 2	Pages 10–12
Explain	Make Your Claim	Day 3	Page 13
	Plants Have Parts	Day 4	Page 14 Video: *What Are Some Parts of Plants*
	Science Read Aloud: Comparing Plants		Page 15
	Close Reading: *Plant Structures*	Day 5	Pages 16–17
Elaborate	Inquiry Activity: *Plant Structures*	Day 6	Pages 18–19
	Trees Are Plants		Pages 20–21
Evaluate	Explain the Phenomenon: How is this sequoia tree different from other plants?	Day 7	Pages 22–24
		7 Days	

Essential Question: What patterns can you find between different plants?

Objective: Students will learn characteristics of common plant structures.

Vocabulary: structure

	FlexTrack A 30 min/day (5 days per week)		FlexTrack B 30 min/day (3 days per week)
Pacing	**Resources**	**Pacing**	**Resources**
Day 1	Page 7 Employ the Volleyball Not Ping Pong strategy and limit discussion time.	Day 1	Page 7 Employ the Volleyball Not Ping Pong strategy and limit discussion time.
	Pages 8–9 Video: *Sequoia National Park*		Pages 8–9 Video: *Sequoia National Park*
Day 2	Pages 10–11 Conduct this investigation as a whole class by viewing photos on the board.	Day 2	Pages 10–11 Conduct this investigation as a whole class by viewing photos on the board.
Day 3	Page 13 Review the evidence collected in the Explore activity as a class. Have students make their own claim.		
Day 4	Page 14 Video: *What Are Some Parts of Plants*		Page 14 Video: *What Are Some Parts of Plants*
Day 5	Page 15	Day 3	
Day 6	Pages 16–17 Omit find evidence.		
Day 7	Pages 22–24 Answer the Explain the Phenomenon question as a class.	Day 4	Pages 22–24 Answer the Explain the Phenomenon question as a class.
7 Days		**4 Days**	

Lesson 1: Plant Parts

ENGAGE EXPLORE EXPLAIN ELABORATE EVALUATE

Lesson Objective

Students will explore plants to identify common plant parts. They will construct explanations and make observations about plant structures.

DCI Structure and Function

LS1.A All organisms have external parts. Different animals use their body parts in different ways to see, hear, grasp objects, protect themselves, move from place to place, and seek, find, and take in food, water and air. Plants also have different parts (roots, stems, leaves, flowers, fruits) that help them survive and grow.

LESSON 1

Plant Parts

8 Module: Plant Structures and Functions

Teacher Toolbox

Science Background

Despite the wide variety of plants on Earth, many patterns exist among plants. Plants must accomplish many of the same basic functions: collect water and nutrients, pass on genetic information, collect sunlight, and make food through photosynthesis. As a result, they have developed many similar structures. Students should begin observing similarities and differences among plants. They should begin using the differences between plants to tell them apart from one another.

DISCOVER
THE PHENOMENON

How is this sequoia tree different from other plants?

 GO ONLINE

Watch the video *Sequoia National Park* to see the phenomenon in action.

Look at the photo. Watch the video.
What parts of a tree do you know?
What do you observe?

Sample answer: I see that this sequoia tree
has a trunk, leaves, and roots. This sequoia
tree is much larger than other plants.

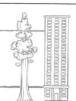

Did You Know?
This tree is very tall. It is as tall as a building with twenty-five floors.

Engage   **Lesson 1** Plant Parts   **9**

 GO ONLINE

INTERACTIVE PRESENTATION

Discover the Phenomenon:
Plant Parts

DISCOVER THE PHENOMENON

 10 min   whole class

Recall that scientists refer to an event or situation that is observed or can be studied as a phenomenon. Have students study the picture of General Sherman.

Ask the **Discover the Phenomenon** question:

How is this sequoia tree different from other plants?

 This leads to the overarching module **Essential Question:** What patterns can you find between different plants?

 GO ONLINE check out *Sequoia National Park* to see the phenomenon in action.

Talk About It

Ask students to describe what they see. Help students turn their observations from the video into questions. Start a class discussion with the following prompts:

ASK: Do you know the name of any of the parts of a tree?

ASK: How is General Sherman similar to other plants you have seen? How is it different?

ASK: What else would you like to learn about General Sherman or other trees?

Record responses and questions on the board or chart paper to refer to as you move through this lesson.

Did You Know?

Explain to students that General Sherman is the largest tree on Earth when measured based on its volume. Students will likely be unfamiliar with the concept of volume. Explain that General Sherman takes up more space than any other living tree on Earth.

General Sherman is also 83.3 meters (275 feet) tall which is as tall as a 25-story building. Even though General Sherman is the largest tree when measured by volume, there are other trees that are taller and more massive than General Sherman. Have students discuss the difference between biggest, tallest, and most massive. Encourage students to conduct research to learn more about record-setting trees.

INQUIRY ACTIVITY | Hands On

Observe Plant Parts

 Prep: 10 min | **Class:** 30 min   small groups

Purpose

Students will observe that plants are made of many separate parts.

Materials

Additional: plastic gloves, batteries

Alternative: If outdoor plants like flowers or trees are available, this activity can be done outside. If live plants cannot be found, photos of plants can be substituted. If photos are used, be sure that a minimum of three parts are visible in each photo. The photos can show the same plant or a variety of plants.

Before You Begin

Have students discuss what they know about plants. Focus students on the question they will investigate:

ASK: How are parts of the plant different?

Post the question to revisit as a class during the activity.

Guide the Activity

Make a Prediction Students will use their prior knowledge to predict the differences between plant parts. If students are struggling to make their prediction, have them look at the photo of the plant on this page to get ideas. Discuss as a class what students predict. When this activity is over, tell students to wash their hands.

Investigate

BE CAREFUL Students should wear non-allergenic gloves when handling plants. If students observe plants outside, make sure the plants are safe to touch in case students touch them without gloves. Bring a first aid kit if students are going outside to make observations.

1. Help students select three distinct parts of the plant such as roots, leaves, stems, or flowers. It is okay if students are not familiar with these terms. Point out structures for students to observe.

5. Shine a flashlight on each plant part. Have students record whether the light passes through each plant part.

Inquiry Spectrum

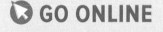 **GO ONLINE** for guidance on how to adapt this activity to a different level of inquiry.

INQUIRY ACTIVITY

Materials
- plant
- hand lens
- crayons
- flashlight

Hands On

Observe Plant Parts

You observed the parts of a tree. Observe parts of another plant.

Make a Prediction How are parts of the plant different?

Sample answer: They have different shapes.

Investigate

BE CAREFUL Wear gloves.

1. Choose three different plant parts.

2. Draw a picture of each part in the table.

3. Use the hand lens. Look at each part. Observe the color and shape.

4. Use your hands. Carefully feel each plant part.

5. Use a flashlight. Shine light on each plant part.

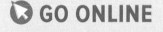 **GO ONLINE**

INTERACTIVE PRESENTATION	ADDITIONAL RESOURCE
Inquiry Actitivy: Observe Plant Parts	Inquiry Rewind: Observe Plant Parts
INQUIRY ACTIVITY	**INQUIRY REWIND**

6. Record Data Write what you observe.

Plant Part	Observations
Drawing could include a stem showing the correct shape and color.	Sample answer: The stem feels stiff and smooth. Light does not shine through the stem.
Drawing could include a leaf showing the correct shape and color.	Sample answer: The leaf feels smooth. Some light shines through the leaf.
Drawing could include roots showing the correct shape and color.	Sample answer: The roots feel bumpy. Light does not shine through the roots.

 Talk About It

Why do you think the light only shines through some plant parts? Tell a partner.

Communicate

Help students determine whether or not their observations matched their predictions. Tell students to discuss how the parts of their plant were similar and different. Connect this activity to the module phenomenon. Ask the class to think about and discuss which plant part they think helps plants stand tall. Encourage students to provide reasoning to help explain their answer.

Waves and Their Applications in Technologies for Information Transfer 1-PS4-3

This activity introduces topics that will be revisited in another module. In the Module *See Objects* students will plan and conduct an investigation to determine the effect of placing objects made with different materials in the path of a beam of light.

 Talk About It

Have students think about why light might shine through some parts of a plant and not others. Tell students to think about what plants need to live and grow.

Short on Time?

If you are short on time, conduct this investigation as a whole class by viewing photos on the board.

SEP **Analyzing and Interpreting Data**

Although Analyzing and Interpreting Data is not assessed in this lesson, students in first grade should be collecting, recording, and sharing observations. Draw attention to the variety of senses and tools that can be used to help make observations.

ASK: What tools did you use to make observations? I used a hand lens, a flashlight, and rubber gloves. I also used my eyes and hands.

Teacher Toolbox

Science Background

In this activity, students will observe how light interacts with various parts of a plant. They may observe that light can pass through some structures, like leaves, and that light is blocked by other structures, like tree trunks. This will help to build prior knowledge or reinforce concepts learned in the Module *See Objects*.

INQUIRY ACTIVITY | Hands On

Observe Plant Parts (continued)

 20 min   whole class

Before Reading

Have students make observations about the picture on page 4 of the Science Read Aloud *Plant Parts Around the World*. Ask students to share what they know about plants from different places around the world. Remind students of what they learned in the Inquiry Activity *Observe Plant Parts*.

💬📖 Science Read Aloud

Read aloud with students pages 4–13 of the Science Read Aloud *Plant Parts Around the World*. Students will learn about plants in a botanical garden. They will be introduced to specific plant structures.

READING❭ Connection

Integration of Knowledge and Ideas - RL.1.7
Ask students to recall the setting of this story. Encourage students to look back in the text to verify their answer before recording it in their student notebook.

After Reading

Ask students to summarize what happened in the reading.

ASK: What did Marco and his dad do? Marco and his dad walked around a botanical garden and learned about plants.

ASK: What did they learn? They learned about different types of plant structures like flowers, roots, leaves, and stems.

INQUIRY ACTIVITY

📖 Read *Plant Parts Around the World*.

7. Do any of Marco's observations about plant parts match what you observed in this activity? Explain using data.

Sample answer: Yes. Marco and I both observed that some plants have leaves and stems.

8. READING❭ Connection Where does this story take place? How do you know?

This story takes place in a botanical garden. The illustrations are of different types of plants.

💬 **Talk About It**

Which part do you think helps the plant stand tall? Tell a partner.

🖱 GO ONLINE

INTERACTIVE PRESENTATION

Read Aloud: Plant Parts Around the World

Differentiated Instruction

AL Record student observations of the page 10 Inquiry and read aloud on a T-chart. Help them use the chart to discuss patterns in the structures of plants.

OL Have students fill out Venn Diagrams to help them notice patterns in the structures of the plant parts they observed in the page 10 Inquiry and read aloud.

BL Have students use the patterns in data they collected on page 11 to make predictions about the plant structures they will find in the read aloud.

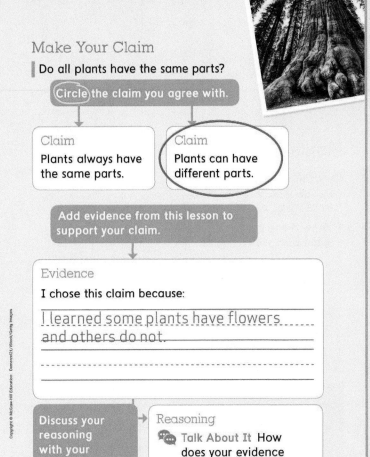

Make Your Claim

Do all plants have the same parts?

Circle the claim you agree with.

Claim	Claim
Plants always have the same parts.	Plants can have different parts.

Add evidence from this lesson to support your claim.

Evidence

I chose this claim because:

I learned some plants have flowers and others do not.

Discuss your reasoning with your classmates.

Reasoning

Talk About It How does your evidence support your claim?

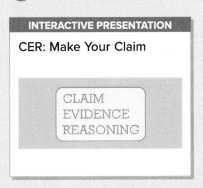 GO ONLINE

INTERACTIVE PRESENTATION

CER: Make Your Claim

CLAIM
EVIDENCE
REASONING

MAKE YOUR CLAIM

 Class: 10 min   whole class

Students will be introduced to the question: Do all plants have the same parts? Students will select the claim they agree with and support their claim with evidence from the Inquiry Activity *Observe Plant Parts*. As the lesson continues, students will revisit this page to add evidence. Lastly, have students provide reasoning to explain how their evidence supports their claim.

CLAIM

Give students time to reflect and brainstorm possible answers. Have students circle the answers in the graphic organizer that they agree with.

EVIDENCE

Scientific evidence is information that supports or contradicts a claim. This information can come from a variety of sources. Research, experimentation, or data interpretation are common sources of scientific evidence. Students can provide evidence based on what they observed in the Inquiry Activity. Throughout the lesson, encourage students to return to their claim to add more evidence. Look for the blue square for a reminder to revisit this page.

If students find that the evidence they collect does not support their claim, they should pause to consider why this occurred. Perhaps an investigation was flawed, or students failed to gather enough information. If this is the case, have students repeat their investigation or continue gathering information. It is also possible that students chose the wrong claim. Give students the chance to correct their claim if they find multiple pieces of evidence that prove their claim is false.

REASONING

When providing reasoning, students must explain the scientific knowledge, principle, or theory they used to support their argument. If, for example, a student claims that plants can have different parts, they must provide the scientific explanation supporting their claim. They may choose to point out text from the Close Reading that states, "Plants can have different structures based on where they are found and what they need to live."

Plants Have Parts

 30 min   whole class

VOCABULARY

Encourage students to use context clues to derive the meaning of the vocabulary words. Struggling students can use the glossary to access the definitions.

GO ONLINE to watch the video *What Are Some Parts of Plants?* to learn more about plant structures.

Visual Literacy

Read a Diagram Have students study the photo on page 14. Tell students to match each part of the plant with the correct description. Quiz students by asking them to identify what plant part is being described.

ASK: What structure has colorful petals? Flowers

ASK: What structure looks like hair and grows in the soil? Roots

Leveled Reader

Use the Leveled Reader *Parts of Plants*. This book describes the parts of plants.

Have students read the book with a partner. After reading, students can discuss how plant parts are similar and different.

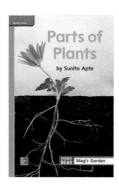

Parts of Plants
by Sunita Apte

Vocabulary

Listen for this word as you learn about plant parts:

structure

Plants Have Parts

GO ONLINE

Watch the video *What Are Some Parts of Plants?* to learn more.

1. Draw a line connecting each plant part with its name.

flower: white and pink with many petals

stem: strong and tall, cylinder shape

leaf: green, feels waxy, curved

root: brown, looks like hair, in the soil

14   Explain   **Module:** Plant Structures and Functions

GO ONLINE

INTERACTIVE PRESENTATION

Read About: Plants Have Parts

ADDITIONAL RESOURCES

Lesson Vocabulary: Plant Parts

vocabulary

Teacher Toolbox

Plantae is one of the six kingdoms of living things. There is a lot of diversity among plants. Some plants, including certain species of algae, are made up of a single cell. Other organisms, including moss and seaweed, do not have vascular systems. These organisms therefore do not have roots. One thing all plants have in common is that they have the organelle, chloroplasts, in their cells. This is the feature that gives plants their greenish color and lets them absorb energy from the Sun.

Listen to *Comparing Plant Parts.*

2. Draw a picture of a plant. Label each part.

Drawing should include a plant with each structure labeled. Labels could include stem, roots, flower, or leaves.

3. How do you think these parts help plants survive?

Sample answer: These parts help plants get water and make food.

🔊 **GO ONLINE**

Explore *Parts of Plants* to see more plant structures.

🔊 GO ONLINE

INTERACTIVE PRESENTATION	**ADDITIONAL RESOURCE**
Read Aloud: Comparing Plant Parts 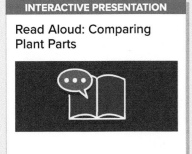	Leveled Reader: Parts of Plants 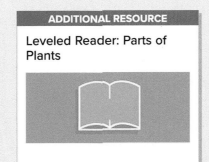

Before Reading

Have students observe the photos on page 14 of the Science Read Aloud *Comparing Plant Parts.* Ask students to share what they know about different plant parts.

🔊 Science Read Aloud

Read aloud with students pages 14–23 of the Science Read Aloud *Comparing Plant Parts.* Students will compare plant parts to find similarities and differences. They will encounter the word *structure.* Discuss the questions on page 24 with students.

After Reading

Have students compare the fiction and nonfiction texts.

READING▶ Connection

Craft and Structure - RL.1.5

Ask how these two texts were similar. Have students discuss the plant structures they observed in each text.

🔊 **GO ONLINE** to explore the Go Further activity *Parts of Plants* to learn more about plant parts.

CCC Patterns

Although this crosscutting concept is not assessed in this lesson, point out that patterns exist everywhere. In first grade, students should be classifying objects. This can be done by looking for similarities and differences between objects. Relate this crosscutting concept to students' observations about the plants they have seen.

ASK: What patterns did you see between the different parts of the plants? Sample answer: I noticed that all of my plant parts were green.

ASK: What patterns did you see between types of plants? Sample answer: I noticed that many plants had tall stems that help up leaves and flowers.

🔊 **GO ONLINE** Have students listen to the Crosscutting Concept Science Song: *Patterns.*

Crosscutting Concept Graphic Organizer

🔊 **GO ONLINE** Use the Crosscutting Concept Graphic Organizer to identify patterns of plant structures.

ENGAGE   EXPLORE   **EXPLAIN**   ELABORATE   EVALUATE

 READING

Plant Structures

 30 min   partners

Inspect

Read Have students read the passage to focus on understanding the overall meaning. Ask students to take notes to gather information about plant structures. Ask students to think about how you can tell plants apart.

ASK: What plant structures are being shown in the photos on pages 16 and 17? Flowers and leaves

ASK: What is a structure? A part of something

Find Evidence

Have students reread the text, looking for answers to the question: How can you tell plants apart?

Introduce students to the skimming strategy. Help students determine what they are being asked to do. Skim line by line to find evidence from the text. Remind students to underline the evidence they find.

Scientific Vocabulary

structure The word structure comes from the Latin word *strure* meaning "to build." Students may be familiar with the word structure when it is used to describe a building. Tell students that the word structure can also be used to describe the parts of a complex object.

Academic Vocabulary

evidence The word evidence has been used previously in this lesson on the Make Your Claim page. Have students observe the word in context. Tell students to look up the word in the dictionary. Develop a classroom definition then look up the word to compare.

A C T **Access Complex Text**

Prior Knowledge Students may not be familiar with the terms: leaves, stems, roots, and flowers. Encourage students to refer to the Explain section of this lesson for examples of these structures. Showing students additional examples from print or online sources will help build context around these terms and help overall comprehension of this text.

ASK: What do flowers look like? Sample answer: Flowers have petals. They can be many different colors.

ASK: What do roots look like? Sample answer: Roots are brown and found in the soil.

 READING

Inspect

Read *Plant Structures.* Circle what the word structures means.

Find Evidence

Underline clues that tell you how you can tell plants apart.

Notes

_____

_____

_____

_____

_____

Plant Structures

Many plants have the same structures. A structure is a part of something. Leaves, stems, roots, and flowers are plant structures. Not all plants have the same structures. Plants can have different structures based on where they are found and what they need to live. A structure may also look different in some plants.

16   Explain   **Module:** Plant Structures and Functions

🔾 GO ONLINE

INTERACTIVE PRESENTATION

Close Reading: Plant Structures

Make Connections

 Talk About It

How might you use leaves or roots to tell plants apart?

Notes

_____

_____

_____

_____

You can use differences in structures to tell plants apart. Some plants have flowers. The flowers can look different. You can use flowers to tell plants apart.

Look at the photos. How are these plant structures different? Explain.

Sample answer: These flowers are a different color and shape.

_____

- - - - - - - - - - - - - - - - - -

_____

Explain   **Lesson 1** Plant Parts   **17**

GO ONLINE

ADDITIONAL RESOURCE

Science Song: Structure and Function

Make Connections

Talk About It

Have partners think about how roots and leaves can be used to tell plants apart. Students should cite evidence from this lesson or their own experiences.

REVISIT
PAGE KEELEY
SCIENCE PROBES Have students return to the Page Keeley Science Probe. Students should revise their answers based on what they have learned during this lesson. Encourage discussion about how students' opinions about plant parts have changed.

CCC Structure and Function

Students have not yet learned the word *function*, however, have students think about how the shape of each plant part might help it serve a purpose.

GO ONLINE Have students listen to the Crosscutting Concept Science Song: *Structure and Function*.

Formative Assessment

Use this opportunity to do a quick assessment to determine whether students are ready to move on. You may choose to assess students as a whole group or individually using the exit slip strategy. Exit slips with the following question can be passed out to students before the end of class. Have students record their responses. Review student responses and provide feedback. Return the corrected exit slips to students the following day.

ASK: What are some structures of plants? Sample answer: Plants have roots, leaves, stems, and flowers.

Lesson 1: **Plant Parts**

INQUIRY ACTIVITY | Hands On

Plant Structures

 Prep: 15 min I **Class:** 30 min   small group

Purpose

Students will compare an onion and a daisy to find which structures they have in common.

Materials

Collect a daisy and an onion plant from a local nursery or supermarket. Make sure these plants include roots and flowers. If you are unable to find an onion plant with a flower, show students a photo of an onion flower. After this experiment has ended, the plants can be displayed in the classroom until they are ready to be composted. Contact local community gardens for information about composting locations or donations.

Additional: plastic gloves

Alternative: If plants cannot be found, printed pictures can be substituted.

Before You Begin

Have students discuss what they know about a daisy and an onion plant. Focus students on the question they will investigate:

ASK: Which structures do a daisy and an onion both have?

Post the question to revisit as a class during the activity.

Guide the Activity

Make a Prediction Help students make an informed prediction based on structures they have seen on plants in their own lives.

Investigate

BE CAREFUL Students should wear gloves when handling plants. If students do not wear gloves, make sure they wash their hands after the activity.

1. Demonstrate how to properly use a hand lens. Allow students to make and record observations about each plant.

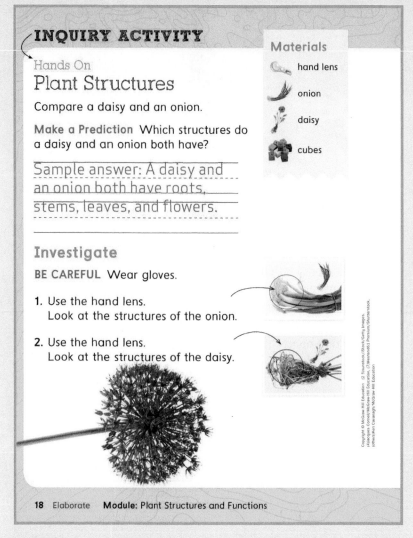

GO ONLINE

INTERACTIVE PRESENTATION
Inquiry Activity: Plant Structures
INQUIRY ACTIVITY

3. Record Data Put a ✔ in the box if the plant has the part.

	Root	Stem	Leaf	Flower
Daisy	✓	✓	✓	✓
Onion	✓	✓	✓	✓

Communicate

4. Does what you learned match your prediction? Explain.

My observations matched my prediction. The daisy and onion had the same parts.

MATH Connection

Use cubes to compare the length of the daisy and the onion. Tell a partner what you find.

Communicate

Have students compare their results with their predictions. Help students determine whether or not their predictions were supported by their observations.

MATH Connection

Measurement and Data   1.MD.A.2

Have students use measuring cubes to compare the size of the onion and the daisy. This will reinforce the math skill of using multiple copies of a shorter object end to end to understand that the length of an object is the number of same-length units that span an object without gaps or overlap. Ask students to count the number of whole cubes and explain which plant is longer.

Remind students of the STEM Career landscape architect. Have students think about how measurements and math skills are related to plants.

ASK: Why might a landscape architect need to know how tall plants grow? Sample answer: A landscape architect needs to know how tall plants grow so that the spaces they make have plants with different heights.

ASK: What could happen if the plants in an outdoor space were taller than the people? Sample answer: People might not be able to see, and they could get lost.

Inquiry Spectrum

⏵ **GO ONLINE** for guidance on how to adapt this activity to a different level of inquiry.

Short on Time?

Do the activity as a whole class by using large images of the structures of an onion plant and a daisy plant.

EL Support

ELD.PII.1.6, LS1.A, LS3.B: Help students connect ideas about plant structures by comparing and contrasting plant traits using a Venn Diagram. Introduce same and different. Point to the non-overlapping area. Ask: Same or different? Different Point to the overlapping area. Ask: Same or different? Same

EMERGING

Compare images of plants as you complete a Venn diagram. Say, Flowers. Same or different? Different Leaves. Same or different? Same

EXPANDING

Have students practice comparing and contrasting using information from the Venn diagram. For example, A fern doesn't have flowers.

BRIDGING

Have students compare or contrast two things in the same sentence. For example, Daisies have flowers, but ferns have spores.

Trees Are Plants

 20 min   whole class

Read the passage and look at the photo with students. Ask students to think about how a trunk is similar to other plant parts they have learned.

CCC **Structure and Function**

Help students identify the structures of plants. Inform students that an object's shape helps it do things. Students will not yet be familiar with the work function but should be introduced to the idea that every object has a purpose or does something. Help students make predictions about the function of various plant structures based on observations.

ASK: Why do you think trees have trunks? Sample answer: I think trees have trunks to hold up the leaves and branches.

ASK: Why do you think daisies have small stems instead of large trunks? Sample answer: A small daisy doesn't need a large trunk to hold it up.

GO ONLINE Have students listen to the Crosscutting Concept Science Song: *Structure and Function*.

■ COLLECT EVIDENCE Have students revisit their claims and add evidence to the Make Your Claim page. Have students discuss their reasoning with the class.

Examples of evidence include quotations from the close reading or observations from an inquiry activity.

Examples of reasoning include quotations from the close reading that state why plants have different parts.

ASK: What is the reason plants have different parts? Sample answer: Plants have different parts based on where they are found and what they need to live.

Teacher Toolbox

Identifying Preconceptions

Students may not realize a tree is a plant. Have students point out the plant structures that they see on the tree. Similarly, students may think a trunk is a new plant structure. In reality, trunks are a type of stem. Although students have not yet learned the functions of plant structures, trunks perform the same function as a stem. It supports the tree and helps move water and nutrients. Each leaf on a tree also has a stem that connects it to a branch. The difference between the stem of the tree (trunk) and the stem on each leaf can be confusing to students.

Trees Are Plants

Not all plants look the same. A tree is a plant. Trees have many structures. They have roots, leaves, and stems. The main stem of a tree is called a trunk.

A trunk helps hold up the tree. A trunk is usually covered with bark. The bark is a hard layer that keeps the tree safe. Trunks can be tall or they can be short. Trunks connect the top of the plant with the bottom of the plant.

20   Elaborate   **Module:** Plant Structures and Functions

GO ONLINE

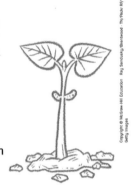

INTERACTIVE PRESENTATION

Read About: Trees are Plants

Ray Sandusky/Brentwood - TN/Moment/Getty Images

1. (Circle) the trunk of the tree in the photo.
2. (Circle) the plant structure that is most like a trunk.

 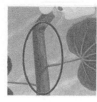 

3. How are the parts of a tree different from the parts of a daisy?

Sample answer: A daisy has a few leaves. Trees have many leaves.

What **structure** do a daisy and a tree both have? How are these **structures** similar?

Sample answer: Both plants have a stem. The stems hold up the plants.

Elaborate **Lesson 1** Plant Parts **21**

Three-Dimensional Thinking

CCC: Structure and Function

DCI: LS1.A Structure and Function

Check that student responses correctly identify the structures a daisy and a tree have in common. Students have not yet learned the function of these structures, but should be identifying similar structures. This will help students build the skills of identifying patterns and making inferences based on observations.

Crosscutting Concept Graphic Organizer

GO ONLINE Use the Crosscutting Concepts Graphic Organizer to see how structure and function apply to the parts of plants.

EL Support

ELD.PII.1.6, SEP-7: Help students make connections between and join ideas they come across during the lesson to formulate an argument about whether plants have the same or different parts and support it with evidence.

EMERGING
Help students record evidence with pictures and phrases on sticky notes. Have them combine the notes using "and, but, so" to form arguments.

EXPANDING
Use sentence frames to help students formulate arguments and evidence statements. Ex: I learned ___ and ___. I observed that ___. In the book/video ___.

BRIDGING
Use sentence strips to make students' arguments more complex by writing down their statements, cutting them apart, and combining them using conjunctions.

LESSON 1 REVIEW

 20 min   whole class

EXPLAIN THE PHENOMENON

Have students revisit the photo of General Sherman. Show the *Sequoia National Park* video again.

| **Rediscover the Phenomenon:**

How is this sequoia tree different from other plants?

This leads to the overarching lesson **Essential Question:**

What patterns can you find between different plants?

Encourage students to review the notes and questions they wrote. Students should try to answer the questions that they had at the beginning of the lesson.

 Students should revisit the Page Keeley Science Probe to decide whether they would like to change or justify their response. Students have had an opportunity to develop a conceptual understanding of plant structures. Revisiting the probe here will reveal whether students are holding on to a misconception or have gaps in conceptual understanding.

 GO ONLINE to explore the Vocabulary Flashcards with students to review lesson vocabulary.

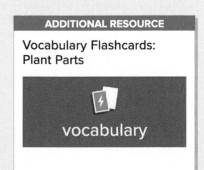

EXPLAIN THE PHENOMENON | How is this sequoia tree different from other plants?

Summarize It

How are the structures of plants similar and different? Use you observations to explain.

Sample answer: All plants have structures. Plants have different parts. Many plants have roots, stems, and leaves. Some plants have flowers.

REVISIT PAGE KEELEY SCIENCE PROBES | Look at the Page Keeley Science Probe on page 7. How has your thinking about plant structures changed?

22 Evaluate   **Module:** Plant Structures and Functions

GO ONLINE

INTERACTIVE PRESENTATION	ADDITIONAL RESOURCE
Lesson Review: Plant Parts	Vocabulary Flashcards: Plant Parts

Three-Dimensional Thinking

Answer these questions based on what you learned about plants.

1. Compare these plants. (Circle) the structures they both have.

2. Which statement is true about plant structures?

a. Flowers can be used to tell plants apart.

b. All plant parts look the same.

c. All leaves are the same shape.

Copyright © McGraw Hill Education (IRadhnaf/Getty Images, (r)Lev Kropotov/Shutterstock.com

Evaluate   **Lesson 1** Plant Parts   **23**

GO ONLINE

ADDITIONAL RESOURCE

Lesson Check: Plant Parts

☑

ADDITIONAL RESOURCES

Vocabulary Check: Plant Parts

☑ vocabulary

LESSON 1 REVIEW

Three-Dimensional Thinking

Have students apply their three-dimensional learning to show their understanding.

1-LS1-1 Use materials to design a solution to a human problem by mimicking how plants and/or animals use their external parts to survive, grow, and meet their needs.

1. Both the flower and the tree have stems and flowers. As described in the Elaborate section of this activity, tree trunks are a type of stem. LS1.A, CCC-6, DOK-2

2. Statement A is correct. Flowers differ from plant to plant and can be used to tell plants apart. SEP-6, LS1.A, DOK-1

Online Assessment Center

You might want to assign students the lesson check that is available in your online resources. You can assign the premade lesson check, which is based on the Disciplinary Core Ideas for the lesson, or you can customize your own lesson check using the customization tool.

GO ONLINE explore the Vocabulary Check: Plant Parts with students to review the vocabulary from the lesson or assign to students to evaluate their lesson vocabulary knowledge.

Differentiated Instruction

AL Give each student a picture of a one plant part. They will move about the room to find classmates with similar plant structures. Then, share out the patterns they find.

OL Ask students to cut up pictures of different plants into parts. Then, sort the parts into similar structures, and discuss the patterns they find with a partner.

BL Challenge students to a scavenger hunt in their books. Ask them to locate words and pictures that show plants have similar parts.

LESSON 1 REVIEW

Extend It

 15 min   individual

This task focuses on the 21st Century Skills of critical thinking as well as problem solving. This open inquiry activity gives students the opportunity to plan an investigation to find the answer to a question. Encourage students to conduct additional research about plants as needed.

Extend It Scoring Rubric

Use the following rubric guidelines to assess the Extend It activity.

4 Points The student has made a list of potential questions, selected one question, planned an investigation, and found the answer to their question (or identified what additional steps would be needed to find an answer).

3 Points The student made a list of potential questions, selected one question, planned an investigation, but failed to find the answer to their question (or identify what additional steps would be needed to find an answer).

2 Points The student made a list of potential questions, selected one question, but did not plan a reasonable investigation, and failed to find the answer to their question (or identify what additional steps would be needed to find an answer).

1 Point The student made a list of potential questions, but did not select a single question, plan an investigation, or find the answer to their question (or identify what additional steps would be needed to find an answer).

Extend It

What questions do you still have about plant parts?

Answers will vary. Sample answer: What would happen if a plant did not have leaves? How are flowers and trees similar? What are the different structures of a cactus?

Write the question you want to investigate.

Answers will vary. Sample answer: What are the different structures of a cactus?

Plan and conduct an investigation to answer your question.

Write the answer to your question.

Answers will vary. Sample answer: A saguaro cactus has roots, a trunk, ribs, fruit, arms, and a crown.

24 Evaluate   **Module:** Plant Structures and Functions

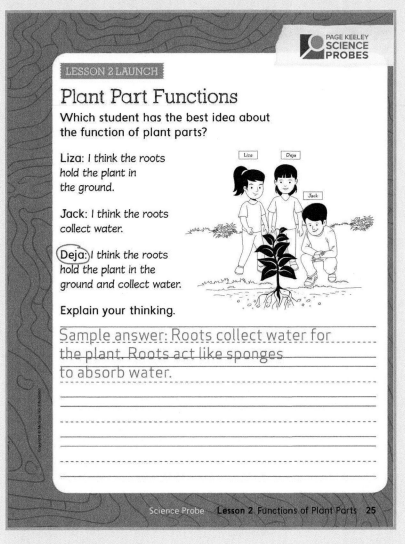

Plant Part Functions

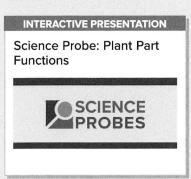 10 min   whole class

Using the Probe

The purpose of this probe is to identify students' prior knowledge about the function of parts of plants. This probe works well as a whole class discussion. Have students think about and mark their answers in their notebook. Then have students share their ideas with the class. Give students time to make changes to their original ideas when they record their explanations. Use this probe to assess students' prior knowledge of the lesson content and to identify possible misconceptions.

Be sure not to tell students the answer. It is not important that students know the answer to this probe at this point in the lesson. What is important is the reasoning students provide to support their answer. Students will revisit this probe throughout the lesson to see how their thinking has changed.

GO ONLINE to learn about other strategies to use with this probe.

Throughout the Lesson

Use students' explanations to bridge the students' initial ideas about the function of plant structures with the understanding they develop. Prompts in the Teacher's Edition will instruct you when it's time for students to revisit this probe.

Teacher Explanation

Deja has the best idea about the function of roots. She correctly understands that roots hold the plant in the ground and collect water.

GO ONLINE

INTERACTIVE PRESENTATION

Science Probe: Plant Part Functions

SCIENCE PROBES

Teacher Toolbox

Identifying Preconceptions

This probe is designed to reveal if students can recognize the function of roots. Students may not realize that roots accomplish multiple functions for the plant. Students may not understand that roots help hold plants in the ground while also collecting water and nutrients from the soil.

Building to the Performance Expectations

In this lesson, students will explore content and develop skills leading to mastery of the following Performance Expectations:

1-ESS1-1. Use observations of the sun, moon, and stars to describe patterns that can be predicted.*

1-LS1-1. Use materials to design a solution to a human problem by mimicking how plants and/or animals use their external parts to help them survive, grow, and meet their needs.

K-2-ETS1-2. Develop a simple sketch, drawing, or physical model to illustrate how the shape of an object helps it function as needed to solve a given problem.

*This Performance Expectation is introduced but not assessed in their entirety during this lesson.

SEP Science and Engineering Practices

Constructing Explanations and Designing Solutions

Constructing explanations and designing solutions in K–2 builds on prior experiences and progresses to the use of evidence and ideas in constructing evidence-based accounts of natural phenomena and designing solutions. (1-LS1-1)

Developing and Using Models

Modeling in K-2 builds on prior experiences and progresses to include using and developing models (i.e., diagrams, drawing, physical replica, diorama, dramatization, or storyboard) that represent concrete events or design solutions. (K-2-ETS1-2)

DCI Disciplinary Core Idea

ETS1.B Developing Possible Solutions

Designs can be conveyed through sketches, drawings, or physical models. These representations are useful in communicating ideas for a problem's solutions to other people. (K-2-ETS1-2)

LS1.A: Structure and Function

All organisms have external parts. Different animals use their body parts in different ways to see, hear, grasp objects, protect themselves, move from place to place, and seek, find, and take in food, water and air. Plants also have different parts (roots, stems, leaves, flowers, fruits) that help them survive and grow. (1-LS1-1)

LS1.D: Information Processing

Animals have body parts that capture and convey different kinds of information needed for growth and survival. Animals respond to these inputs with behaviors that help them survive. Plants also respond to some external inputs. (1-LS1-1)

CCC Crosscutting Concept

Connections to Engineering, Technology, and Applications of Science Influence of Science, Engineering and Technology on Society and the Natural World

Every human-made product is designed by applying some knowledge of the natural world and is built using materials derived from the natural world. (1-LS1-1)

Structure and Function

The shape and stability of structures of natural and designed objects are related to their function(s). (1-LS1-1, K-2-ETS1-2)

Writing Connections

W.1.8 Research to Build and Present Knowledge

Track Your Progress to the Performance Expectations

You may want to return after completing the lesson to note concepts that will need additional review before your students start the module Performance Project.

Dimension	Concepts to Review Before Assessment
SEP Constructing Explanations and Designing Solutions (1-LS1-1)	
SEP Developing and Using Models (K-2-ETS1-2)	
DCI ETS1.B Developing Possible Solutions (K-2-ETS1-2)	
DCI LS1.A: Structure and Function (1-LS1-1)	
DCI LS1.D: Information Processing (1-LS1-1)	
CCC Structure and Function (1-LS1-1, K-2-ETS1-2)	
CCC *Connections to Engineering, Technology, and Applications of Science* Influence of Science, Engineering and Technology on Society and the Natural World (1-LS1-1)	

Lesson at a Glance

Full Track is the recommended path for the complete lesson experience. FlexTrack A and FlexTrack B provide timesaving strategies and alternatives.

		Full Track 45 min/day (full year)	
	Day-to-Day	**Pacing**	**Resources**
Assess Prior Knowledge	Page Keeley Science Probe: *Plant Parts Functions*	Day 1	Page 25
Engage	Discover the Phenomenon: Where is the Sun in this photo?		Pages 26–27 Video: *Leaves Move*
Explore	Inquiry Activity: *Plants and Light*	Day 2	Pages 28–30
	Science Read Aloud: *Which Way to Sprout*		Page 31
Explain	Plant Parts and Their Functions	Day 3	Pages 32–33 Video: *How Plants Use Their Parts to Live and Grow*
	Plants Use Their Parts to Live and Grow		Pages 34–35
Elaborate	STEM Career Connection: *What Does A Botanist Do?*	Day 4	Pages 36–37
	Inquiry Activity: *Celery Stems*	Day 5	Pages 38–39
Evaluate	Explain the Phenomenon: Where is the Sun in this photo?	Day 6	Pages 40–42
		6 Days	

Essential Question: What patterns can you find between different plants?

Objective: Students will learn characteristics of common plant structures

Vocabulary: structures

FlexTrack A
30 min/day (5 days per week)

Pacing	Resources
Day 1	Page 25 Employ the Fingers Under Chin strategy.
	Pages 26–27 Video: *Leaves Move*
Day 2	Pages 28–30 Show photos of a plant in the morning and in the afternoon for 3 days.
Day 3	Page 31
Day 4	Page 32 Video: *How Plants Use Their Parts to Live and Grow*
Day 5	Page 33 Video: *Flowers, Fruits, and More*
Day 6	Pages 40–42 Answer the Explain the Phenomenon question as a class.

6 Days

FlexTrack B
30 min/day (3 days per week)

Pacing	Resources
Day 1	Page 25 Employ the Fingers Under Chin strategy.
	Pages 26–27 Video: *Leaves Move*
Day 2	Pages 28–30 Show photos of a plant in the morning and in the afternoon for 3 days.
Day 3	Pages 32–33 Video: *How Plants Use Their Parts to Live and Grow* Video: *Flowers, Fruits, and More* Omit question 1.
Day 4	Pages 40–42 Answer the Explain the Phenomenon question as a class.

4 Days

Lesson Objective

Students will explore plants to determine the function of plant parts. They will construct explanations and build models based on the structure and function of plant parts.

DCI Structure and Function

LS1.A All organisms have external parts. Different animals use their body parts in different ways to see, hear, grasp objects, protect themselves, move from place to place, and seek, find, and take in food, water and air. Plants also have different parts (roots, stems, leaves, flowers, fruits) that help them survive and grow.

DCI Information Processing

LS1.D Animals have body parts that capture and convey different kinds of information needed for growth and survival. Animals respond to these inputs with behaviors that help them survive. Plants also respond to some external inputs.

DCI Developing Possible Solutions

ETS1.B Designs can be conveyed through sketches, drawings, or physical models. These representations are useful in communicating ideas for a problem's solutions to other people.

LESSON 2

Functions of Plant Parts

26 | Module: Plant Structures and Functions

Teacher Toolbox

Science Background

Most plants have the same basic structures like roots, stems, leaves, and flowers that function to keep the plant alive and healthy. In this lesson, students will learn the functions of some plant structures. Throughout the lesson, help students make connections to understand how the shape of a structure helps it accomplish its function.

DISCOVER
THE PHENOMENON

Where is the Sun in this photo?

 GO ONLINE

Watch the video *Leaves Move* to see the phenomenon in action.

Look at the photo. Watch the video. Why do you think the leaves in the video move? What do you observe?

Sample answer: The leaves of the plant follow the light. I think the Sun must be to the left of the photo.

Did You Know?

Some plants will close or curl up at night!

Engage **Lesson 2** Functions of Plant Parts **27**

GO ONLINE

INTERACTIVE PRESENTATION

Discover the Phenomenon: Functions of Plant Parts

DISCOVER THE PHENOMENON

 10 min — whole class

Recall that scientists refer to an event or situation that is observed or can be studied as a phenomenon.

Have students study the photo of the plant following the light.

Ask the **Discover the Phenomenon** question:

Where is the Sun in this photo?

This leads to the overarching lesson **Essential Question**: What do plant structures do?

GO ONLINE check out *Leaves Move* to see the phenomenon in action.

Talk About It

Ask students to describe what they see. Help students turn their observations from the video into questions. Start a class discussion with the following prompts:

ASK: Where do you think the Sun is in this photo?

ASK: What time of day do you think it is?

ASK: Why do you think the plant moves?

Record responses and questions on the board or chart paper to refer to as you move through this lesson.

Did You Know?

Plants have various reactions to sunlight. Some plants, follow the Sun. Others plants like tulips, hibiscus, and poppies have flowers that curl up at night. Students will learn more about plant responses in the Module *Plant Parents and Their Offspring*.

ENGAGE **EXPLORE** EXPLAIN ELABORATE EVALUATE

INQUIRY ACTIVITY | Hands On

Plants and Light

 Prep: 5 min | **Class:** 30 min small group

Purpose

Students will observe how a plant moves in response to the Sun.

Materials

Purchase fully grown mint plants from a local nursery or supermarket. Each group should have its own plant. If necessary, the entire class could share a single plant.

Additional: plastic gloves

Alternative: If a mint plant cannot be obtained, sunflowers, poppies, marigolds, daisies, alfalfa, soybeans, or bean plants can be substituted.

Before You Begin

Identify a location in the classroom, school, or outdoors that receives sunlight throughout the day. Ask students if plants move. Have them discuss their ideas. Focus students on the question they will investigate:

ASK: What will happen to the leaves of a plant as the Sun moves?

Post the question to revisit as a class during the activity.

Guide the Activity

Make a Prediction Students should base their predictions on their prior knowledge and observations of the phenomenon.

Investigate

BE CAREFUL Students should wear non-allergenic gloves when handling plants.

1. Results will be best if the window faces south and receives sunlight throughout the day.

2. Continue observing the plant for three days to confirm results. Students should notice a pattern.

Academic Vocabulary

prediction Students have encountered the word prediction in previous inquiry activities. Encourage students to use their prior knowledge to determine what prediction means on this page. Have students develop a class definition then look up the meaning on the word. Compare the class definition with the official definition.

INQUIRY ACTIVITY

Hands On

Plants and Light

Materials
- plant
- crayons

You observed how the leaves of a plant follow the light across the sky. Investigate what other plants do.

Make a Prediction What will happen to the leaves of a plant as the Sun moves?

Sample answer: The leaves will follow the Sun.

Investigate

BE CAREFUL Wear gloves.

1. Put a plant in a sunny place.

2. Observe the plant in the morning.

3. Observe the plant in the afternoon.

4. **Record Data** Use pictures and words to record your observations.

28 Explore **Module: Plant Structures and Functions**

🔄 GO ONLINE

INTERACTIVE PRESENTATION	ADDITIONAL RESOURCE
Inquiry Activity: Plants and Light	Inquiry Rewind: Plants and Light
INQUIRY ACTIVITY	◄ INQUIRY REWIND

5. Observe the plant for three days.
Add details.

Plant in the morning	Plant in the afternoon
Each drawing should show how the plant responds to the sunlight.	

Earth's Place in the Universe 1-ESS1-1

This activity introduces topics that will be revisited in another module. In Module *Observe the Sky* students will observe and record data about the patterns of movement of the Sun, Moon, and stars. Help students make connections between the movement of the Sun and plant responses. These patterns can be used to make predictions about the future.

CCC Patterns

Although Patterns are not assessed in this lesson, point out that patterns exist everywhere. In first grade, students should recognize the cycling of day and night. They should notice how the natural world responds to these changes. Draw attention to the importance of repeating an experiment in order to confirm results and check for error. In this investigation, data is being collected over three days to ensure accurate results and to better observe the pattern of plant movement.

ASK: What patterns did you notice about the movement of the plant? The plant moved to follow the Sun as it moved across the sky.

ASK: How did the data you collect each day compare? Sample answer: My data were the same the first two days, but on the third day it was dark and cloudy outside so I did not observe the plant moving.

SEP Constructing Explanations and Designing Solutions

Students in first grade should be using their observations to construct explanations about the phenomena they observe. These explanations will become the basis of future investigations to understand phenomena. These explanations can also be used to design devices to solve problems.

ASK: What part of the plant do you think follows the Sun? Sample answer: I think the flowers and leaves of some plants follow the Sun.

Teacher Toolbox

Science Background

In this activity, students will observe the movement of a plant over several days to see a pattern over time. They should observe that the Sun moves across the sky each day and that the plant adjusts to follow the Sun. This will help build prior knowledge or reinforce concepts learned in Module *Observe the Sky*.

Lesson 2: Functions of Plant Parts

ENGAGE **EXPLORE** EXPLAIN ELABORATE EVALUATE

Communicate

Have students compare their observations with their predictions. Encourage students to point out any patterns they observed and explanations they have developed.

 Talk About It

Have partners discuss why they think the leaves of a plant move to face the Sun. Encourage them to explain the reasoning behind their idea.

Inquiry Spectrum

Structured Inquiry

In this Inquiry Activity, students are given a question to investigate and a procedure to follow.

Guided Inquiry

To make this a guided activity, present students with the question. Have them predict what will happen to the leaves of a plant as the Sun moves. Let students work in groups to develop a procedure to answer the question.

Open Inquiry

Present students with the phenomenon video of a plant following the Sun. Present the class with another plant and give students the opportunity to determine questions they would like to investigate. Have students design their own investigation to answer the question they developed.

🌐 **GO ONLINE** Have students listen to the Crosscutting Concept Science Song: *Patterns*

Crosscutting Concept Graphic Organizer

🌐 **GO ONLINE** Use the Crosscutting Concept Graphic Organizer to identify patterns between the direction of a light source and plant movement.

Differentiated Instruction

AL Have students use props or their bodies to act out the pattern they see in their data chart. Have them construct an explanation of what they observed while they act.

OL Help students use their data charts to construct an argument that plant leaves follow the sun with sentence prompts: My data table shows that___. The pattern is ___.

BL Have students make predictions about what causes the leaf pattern. Have them re-visit their predictions as they come across explanations in the chapter.

INQUIRY ACTIVITY

Communicate

6. Did what you observe match your prediction? Explain

Sample answer: Yes. My observation matched my prediction. The leaves changed direction during the day to follow the Sun.

7. How do you think the plant will move on the fourth day? Explain your reasoning.

Sample answer: The leaves will continue to follow the Sun. This is a pattern that will continue to repeat.

 Talk About It

Why do you think the leaves of some plants follow the Sun? Tell a partner.

You observed how the leaves of a plant follow the Sun. The leaves respond to the sunlight.

 Listen to *Which Way to Sprout*.

8. How do the seeds know which way their roots should grow?

Sample answer: Their roots follow the water down.

9. Draw a picture to show how seeds grow. Use arrows to show which way the plant parts grow.

Drawing should include the roots growing downward and the stem growing upward.

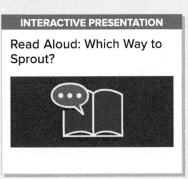 GO ONLINE

Read Aloud: Which Way to Sprout?

INQUIRY ACTIVITY | Hands On

Plants and Light (continued)

 20 min   whole class

Before Reading

Have students observe the picture on page 4 of the Science Read Aloud *Which Way to Sprout?* Ask students to share what they know about how plants grow. Remind students of what they observed in the Inquiry Activity *Plants and Light*.

Science Read Aloud

Read aloud with students pages 4–13 of the Science Read Aloud *Which Way to Sprout?* Students will learn one function of a seed.

After Reading

Ask students to summarize what happened in the reading. Point out that students already observed that plants respond to the Sun. Encourage students to draw connections between the reading and their inquiry activity.

ASK: Besides the Sun, what other things in nature do plants respond to? Plants respond to the direction rain falls.

ASK: How could you design an experiment to test which direction seeds grow? Sample answer: We could plant seeds in a plastic bag so we could see the roots grow.

Differentiated Instruction

AL Before reading, do a picture walk of the read aloud. Have students to ask questions that might be answered in the book. Revisit the questions after reading.

OL Think aloud as you read: The seeds need to solve the problem of which way to grow. How they will know? Stop at key pages and have students practice asking questions and defining problems on sticky notes.

BL Have students write their own short stories about how plants solve a problem in order to get something they need to grow or survive.

Plant Parts and Their Functions

 25 min   whole class

VOCABULARY

Encourage students to use context clues to derive the meaning of the vocabulary words. Struggling students can use the glossary to access the definitions.

Scientific Vocabulary

function   The word structure comes from the Latin word *fungi* meaning "to perform." An object's function describes its purpose or the action it performs.

Before Reading

Have students observe the photo on page 14 of the Science Read Aloud *How Plants Use Their Parts to Live and Grow*. Ask students to predict the function of the plant parts they have learned.

Science Read Aloud

Read aloud with students pages 14-23 of the Science Read Aloud. Students will learn the function of plant parts. While reading, students will encounter the vocabulary words *flower, fruit, leaf, root, seed, and stem*. Discuss the questions on page 24 as a class.

After Reading

Have students compare the fiction and nonfiction texts. Ask how these two texts were similar. Have students discuss the plant structures and functions they observed in each text.

Time to Move

Instruct students to pick a plant structure. Based on what they learned about plant functions in the video, have students act out a plant function. Turn this into a game by having one student perform at a time. Encourage students to raise their hands and guess what plant structure is being shown.

Teacher Toolbox

Science Background

Each part of a plant serves an important function that helps keep the plant healthy. The shape, size, and structure of each part helps it accomplish its purpose. The size of a leaf, width of a stem, and color of a flower help each structure function to capture light, move nutrients and water, or entice pollinators respectively.

Vocabulary

Listen for these words as you learn about what plant parts do.

flower   fruit   function   leaf

root   seed   stem

Plant Parts and Their Functions

 Listen to *How Plants Use Their Parts to Live and Grow*.

A **function** is the purpose of something. Each plant structures has a function. The function describes what each structure does.

1. Draw a picture that shows the function of a root.

> Drawing should inculde a picture of a plant with roots. The roots should be holding the plant in place.

Time to Move

Act out the function of a plant structure.

32 Explain   **Module:** Plant Structures and Functions

GO ONLINE

INTERACTIVE PRESENTATION
Lesson Vocabulary: Function of Plant Parts

vocabulary

INTERACTIVE PRESENTATION
Read Aloud: How Plants Use Their Parts to Live and Grow

INTERACTIVE PRESENTATION
Read About: Plant Parts and Their Functions

Ingram Publishing

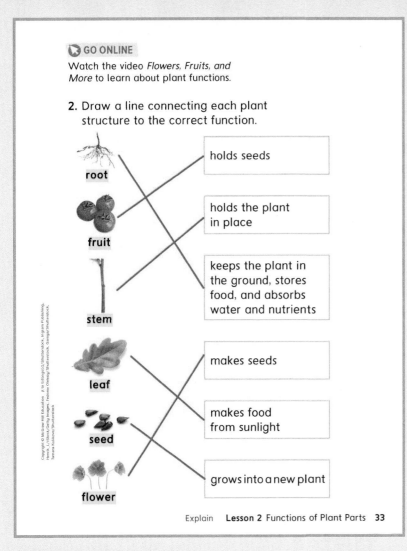
GO ONLINE

ADDITIONAL RESOURCES

Science Song: Structure and Function

Have students **GO ONLINE** to watch the video *Flowers, Fruits, and More* individually or play the video for the whole class to learn about the function of plant structures.

Visual Literacy

Read a Diagram Have students study the photos on this page. Tell students to match each part of the plant with its correct function.

CCC Structure and Function

Have students think about how the shape of each structure helps it function. Encourage students to use the photos to make observations about each structure. Then ask students to think about how the shape of each part helps its function.

GO ONLINE Have students listen to the Crosscutting Concept Science Song: *Structure and Function*.

Formative Assessment

Use this opportunity to do a quick assessment to determine whether students are ready to move on. You may choose to assess students as a whole group or individually using the exit slip strategy.

ASK: What is the function of a leaf? A leaf takes in air and sunlight to make food.

EL Support

ELD.PII.1.6, SEP-6, LS1.A: Help students construct explanations of the structure and function of plants by connecting their ideas about plant parts and the information in the Read Aloud.

EMERGING
After completing the matching activity, have students combine their answers to construct sentences while explaining the functions of plant parts to the group.

EXPANDING
After reading, show key pages of the book and have students use the pictures on each page to explain the function of plant parts to a partner.

BRIDGING
Provide linking words (and, also, by, from) to help students expand their thinking while they explain the functions of different plant parts to a partner.

CLOSE READING

Plants Use Their Parts to Live and Grow

 30 min   partners

Inspect

Read Have students read the passage to focus on understanding the overall meaning. Ask students to take notes to gather information about functions of plant structures. Tell students to think about how plants use their structures to live and grow.

Find Evidence

Have students reread the text, looking for evidence that answers the question: Can plants live without their roots, stems, and leaves?

Remind students to underline the evidence they find.

ASK: What could happen if a plant did not have roots?
The plant might not be able to collect water.

ASK: What could happen if a plant did not have a stem?
The plant might not be able to stay upright.

ASK: What could happen if a plant did not have leaves?
The plant might not be able to make food.

 GO ONLINE to use the digital interactive *How Parts Help a Plant* to learn more about plant part functions.

VKV Visual Kinesthetic Vocabulary

 15 min   whole class

Have students cut out and fill in the Dinah Zike Visual Vocabulary from pages VKV1-VKV2 in their notebook. Guide students in using the VKV to learn the words *leaf*, *flower*, *stem*, and *root*.

Crosscutting Concept Graphic Organizer

 GO ONLINE to have students use the Crosscutting Concepts Graphic Organizer to use throughout the lesson to understand Structure and Function and how it applies to the parts of plants.

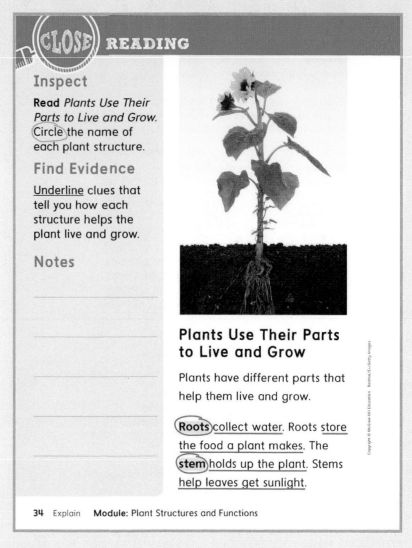

CLOSE READING

Inspect

Read *Plants Use Their Parts to Live and Grow.* Circle the name of each plant structure.

Find Evidence

Underline clues that tell you how each structure helps the plant live and grow.

Notes

_____

_____

_____

_____

_____

Plants Use Their Parts to Live and Grow

Plants have different parts that help them live and grow.

Roots collect water. Roots store the food a plant makes. The stem holds up the plant. Stems help leaves get sunlight.

34   Explain   **Module: Plant Structures and Functions**

GO ONLINE

INTERACTIVE PRESENTATION

Close Read: Plants Use Their Parts to Live and Grow

CLOSE READING

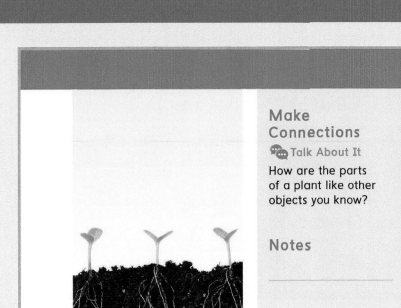

Make Connections

 Talk About It

How are the parts of a plant like other objects you know?

Notes

_____

_____

_____

_____

_____

_____

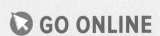

Plants need sunlight to grow. Water and minerals travel through the stem. Leaves take in air and sunlight to make food. A plant gets what it needs to live and grow from its parts.

Explain   **Lesson 2** Functions of Plant Parts   **35**

🔾 GO ONLINE

Make Connections

 Talk About It

Have partners think about the plant structures and functions they have learned. Encourage students to draw connections between plants and other nonliving objects that they are familiar with.

CCC Influence of Science, Engineering and Technology on Society and the Natural World

Inform students that human-made products are designed by applying knowledge of the natural world and using materials derived from the natural world. Have students make a list of products that remind them of things from nature. Encourage students to explain how the objects are similar to nature or the materials from nature that are used to make the objects. Ask the questions below to get students thinking.

ASK: What object do roots remind you of? Why? Sample answer: Roots remind me of a sponge because they both absorb water.

ASK: What object do leaves remind you of? Why? Sample answer: Leaves remind me of a solar panel because they both collect sunlight.

ASK: What is an object you use that is made from things in nature? Sample answer: I use paper which comes from trees.

REVISIT

PAGE KEELEY SCIENCE PROBES

Have students return to the Science Probe. Students should revise their answer based on what they have learned during this lesson. Encourage discussion about how students' opinions about the function of roots have changed.

FOLDABLES **Study Guide Foldables®**

🕐 10 min   👥 whole class

Have students make a Three-Tab Foldables using page EM 49 for guidance. Instruct students to draw a plant structure on the top of each of the tabs. Have students lift each tab and write the name of each structure they drew. Then have students use vocabulary from the lessons to add details about the function of each structure under each tab.

Lesson 2   **Functions of Plant Parts   35**

STEM CAREER Connections
What Does a Botanist Do?

 10 min   whole class

Introduce the botanist STEM Career Connection. Encourage students to look at the photos and share their observations. Have students read question 1 and discuss their answers with the class.

ASK: Why is it important for botanists to learn about plant functions? Sample answer: Botanists need to know how plants parts function to know if plants are healthy or sick.

ASK: What do botanists and landscape architects have in common? Sample answer: Botanists and landscape architects both learn about and work with plants.

Have students answer question 2 and share their drawings with classmates.

PRIMARY SOURCE

Baron Alexander von Humbolt was a Prussian explorer and naturalist. He contributed to modern science in a variety of ways. He conducted extensive mapping of land and water, uncovered ocean currents, studied weather, recorded and collected plant and animal specimens throughout his travels, recorded astronomical data, and was one of the modern era's first environmentalists. Von Humbolt understood the importance of preserving plants, animals, and historical places and worked to keep the natural world safe for future generations. He was also an inspiration to future scientists, including Charles Darwin.

STEM CAREER Connection
What Does a Botanist Do?

Botanists study how plants grow. They study the structure of plants. Botanists learn how plants grow in different environments. Botanists look at plants all over the world.

Baron Alexander von Humboldt was a scientist. He studied plants. He traveled to South America in 1799. He learned about plants that grew in South America.

PRIMARY SOURCE

Baron Alexander von Humboldt

Copyright © McGraw Hill Education. (t)Deco/Kata_(b)Stock/Getty Images; (c)inteksastbastofrou/iStock/Getty Images; (b)Library of Congress Prints and Photographs Division [LC-USZ62-13208B]

36   Elaborate   **Module:** Plant Structures and Functions

GO ONLINE

INTERACTIVE PRESENTATION

STEM Career: Botanist

STEM
CAREER

1. Draw a picture of something people make from plants.

Drawing could include paper, wood, clothing, or food products.

ENVIRONMENTAL Connection

2. Think about what humans need to live. Why is it important that people have healthy plants?

People need healthy plants to eat. Some people also eat animals that need healthy plants. Without healthy plants, there would be no animals to eat.

Have students brainstorm things that humans get from nature. Record a list on the board or chart paper. Encourage students to discuss how they think the relationship between humans and the environment will change as more and more people are born.

ASK: What are some things people get from nature? Sample answer: People get food and water from nature.

ASK: What are some things people make using products from nature? Sample answer: People make paper and clothing out of plants like trees and cotton. People also make oil out of old plants and animals.

ASK: What could happen to nature if more and more poeople are born each year? Sample answer: It might get harder to get the things we need from nature.

ASK: What might happen if people could no longer get what they needed from nature? Sample answer: If people couldn't get food, they would be hungry.

INQUIRY ACTIVITY | Hands On

Celery Stems

 Prep: 10 min | **Class:** 30 min   small group

Purpose

Students will observe how water moves through celery to verify the function of a stem.

Materials

Gather five to six celery stems from a local supermarket. These celery stems must have leaves so students will be able to see the food dye traveling into the leaves.

Additional: plastic gloves

Before You Begin

Focus students on the question they will investigate:

ASK: What is the function of a stem?

Post the question to revisit as a class during the activity.

Guide the Activity

State the Claim Help students select the claim(s) that they believe are true about the function of a stem.

Collect Evidence

1. The more food coloring that is added to the water, the easier it will be to see changes in the celery.

 GO ONLINE Watch the video *Celery Stems* to see a demonstration of this inquiry activity. If students were absent, this video may also be used to help them observe how water travels through a celery stem.

Communicate

Have students compare their results with the claims they wrote. Have students practice explaining to a partner how this activity supported what they know about the function of a stem.

Three-Dimensional Thinking

SEP: Constructing Explanations and Designing Solutions

DCI: LS1.A Structure and Function

Check student explanations for accuracy. Students should be able to relate the movement of the colorful water with a stem's role in transporting water and nutrients from the soil to all parts of the plant.

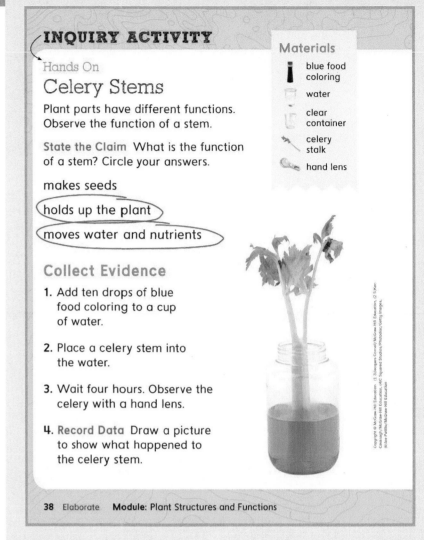

INQUIRY ACTIVITY

Hands On

Celery Stems

Plant parts have different functions. Observe the function of a stem.

State the Claim What is the function of a stem? Circle your answers.

makes seeds

holds up the plant

moves water and nutrients

Materials
- blue food coloring
- water
- clear container
- celery stalk
- hand lens

Collect Evidence

1. Add ten drops of blue food coloring to a cup of water.

2. Place a celery stem into the water.

3. Wait four hours. Observe the celery with a hand lens.

4. **Record Data** Draw a picture to show what happened to the celery stem.

38   Elaborate   **Module:** Plant Structures and Functions

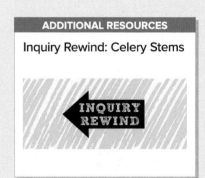 GO ONLINE

INTERACTIVE PRESENTATION	ADDITIONAL RESOURCES
Inquiry Activity: Celery Stems	Inquiry Rewind: Celery Stems
INQUIRY ACTIVITY	**INQUIRY REWIND**

Drawing should include a celery stem in which you can see some blue in the leaves and in the stem.

Communicate

 Explain whether this activity matched what you have learned about the **structure** and **function** of a stem?

Sample answer: The food coloring moved from the water, up the stem, then to the leaves. This shows that the stem carries things from the bottom of a plant to the top.

 Talk About It

How could you design an inquiry activity to show the function of a seed? Tell a partner.

💬 Talk About It

Have students discuss how they could perform an experiment to show the function of a seed. First, have students share their ideas with a partner, then discuss as a class. Keep notes on the board or chart paper. Students may suggest planting a seed in a clear container to see what happens.

Inquiry Spectrum

Confirmation Inquiry

This activity is designed to confirm what students have already learned about the function of a stem.

Structured Inquiry

To make this activity structured, do it before teaching about the function of a stem. Use the same question and procedure given.

Guided Inquiry

To make this a guided inquiry activity, it must be done before teaching about the function of a stem. Provide students with the question: What is the function of a stem? Have students develop their own procedures to answer the question.

Short on Time?

You may wish to do this activity in advance and then show students the results of the activity.

EL Support

ELD.PII.1.6, LS1.A, SEP-6: While asking students to construct explanations about the function of the celery stem, use native English speakers as models for English learners. Have proficient students answer by combining ideas so students will have several models before being required to answer.

EMERGING

Have students point to the drawing and report about what happened during the inquiry. Have them use single words such as *water, stem, up, blue* and so on.

EXPANDING

Ask: What happened to the celery stem? Have students answer using an extended sentence frame: The stem _____. The functions of the stem are to ____.

BRIDGING

Prompt students to look back at the inquiry page to notice some elements they should include in an inquiry activity (materials, steps, recording data, etc.). Then, have them discuss the Talk About It.

Lesson 2: Functions of Plant Parts

LESSON 2 REVIEW

 20 min   whole class

EXPLAIN THE PHENOMENON

| Ask the **Discover the Phenomenon** question:
Where is the Sun in this photo?

 This leads to the overarching module
Essential Question: What do plant structures do?

Have students revisit the photo of the plant as they answer the Explain the Phenomenon question. You may want to show the *Leaves Move* video again.

Encourage students to review the notes and questions they wrote. Students should try to answer the questions that they had at the beginning of the lesson.

 Students should revisit the Science Probe to decide whether they would like to change or justify their response. Students have had an opportunity to develop a conceptual understanding of the function of plant structures. Revisiting the probe here will reveal whether students are holding on to a misconception or have gaps in their conceptual understanding.

LESSON 2

Review

EXPLAIN THE PHENOMENON | Where the Sun is in this photo?

Summarize It

Draw a picture of a plant. Label the function of each plant structure in your picture.

> Drawing should include all plant structures. Labels should describe the function of each structure.

REVISIT SCIENCE PROBES — Look at the Science Probe on page 25. Has your thinking about the function of plant parts changed?

40   Evaluate   **Module:** Plant Structures and Functions

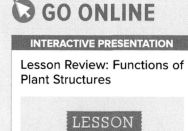 GO ONLINE

INTERACTIVE PRESENTATION	ADDITIONAL RESOURCES
Lesson Review: Functions of Plant Structures	Vocabulary Flashcards: Functions of Plant Structures

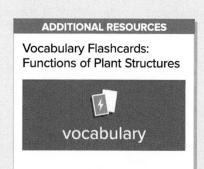

ELD English Language Support

ELD.PII.1.6, LS1.A: Help students connect their ideas about the investigation using the words structure and function. As with many English words, these terms have multiple meanings. Students should focus on how something is built (structure) and what something does (function).

EMERGING
Ask yes/no questions, such as, Is it part of the structure? yes Is a stem a function? no Does a stem move water in a plant? yes

EXPANDING
Say, Draw the structure of a flower. Then ask, What is the function of a stem? Encourage the complete sentence: A stem moves water.

BRIDGING
Have students select an element of a plant's structure and describe its function. A stem is part of the structure of a flower. It moves water through the plant.

Three-Dimensional Thinking

Use what you have learned to answer the questions.

1. Which of the following sentences explains the problem in the picture?

   a. The leaves on the plant in the shade need sunlight.

   b. The roots on the plant in the shade need sunlight.

 Now that you're done with the lesson, answer these questions.

2. A student wants to test the function of a carnation stem. Put the steps in order.

   [3] Wait four hours and observe.

   [2] Put the carnation in the colorful water.

   [1] Add food coloring to the water.

Evaluate **Lesson 2** Functions of Plant Parts **41**

🌟 GO ONLINE

ADDITIONAL RESOURCES	**ADDITIONAL RESOURCES**
Lesson Check: Functions of Plant Parts	Vocabulary Check: Functions of Plant Parts
✓	✓ vocabulary

LESSON 2 REVIEW

Three-Dimensional Thinking

Have students apply their three-dimensional learning to show their understanding.

1-LS1-1 Use materials to design a solution to a human problem by mimicking how plants and/or animals use their external parts to survive, grow, and meet their needs.*

K-2-ETS1-2 Develop a simple sketch, drawing, or physical model to illustrate how the shape of an object helps it function as needed to solve a given problem.

1. A is correct. The leaves on the plant in the shade cannot get enough sunlight. Therefore the plant cannot survive. SEP-6, LS1.D, DOK-3

2. The statements should be labeled 3, 2, 1 respectively. Students must first put food coloring in water, place the carnation stem in the colorful water, then wait to make observations. LS1.A, CCC-6, DOK-2

Online Assessment Center

You might want to assign students the lesson check that is available in your online resources. You can assign the premade lesson check, which is based on the Disciplinary Core Ideas for the lesson, or you can customize your own lesson check using the customization tool.

🌟 **GO ONLINE** explore the Vocabulary Check: Function of Plant Parts with students to review the vocabulary from the lesson or assign to students to evaluate their lesson vocabulary knowledge.

LESSON 2 REVIEW

Extend It

 15 min   Individual

This task focuses on the 21st Century Skills of creative thinking and problem solving. This open inquiry activity gives students the opportunity to solve a problem. Collect print or digital resources that students can use to conduct additional research if needed.

WRITING ⟩ **Connection**

Research to Build Present Knowledge - W.1.8
Help students identify Rishi's problem. Help them solve this human problem by using what they have learned about plant structures and functions. Encourage students to conduct additional research if needed to help Rishi. Student designs should be based on plant structures.

Extend It Scoring Rubric

Use the following rubric guidelines to assess the Extend It activity.

3 Points The student correctly states the function of a plant structure and provides a plausible explanation to solve the problem.

2 Points The student either correctly states the function of plant structure or provides a plausible explanation to solve the problem, but does not accomplish both.

1 Point The student attempted an answer but does not correctly state the function of plant structure and fails to provide a plausible explanation to solve the problem.

Extend It

WRITING Connection Rishi wants to put a sign in his yard. The sign keeps falling over. Use what you have learned about plants. Explain how to solve this problem. Use pictures and words.

Drawing could include a structure similar to roots that hold the sign in the ground.

Sample answer: I thought about roots. I put wires in the ground. The wires hold the sign in place just like roots hold a plant in palce.

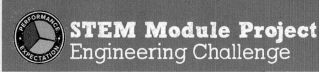

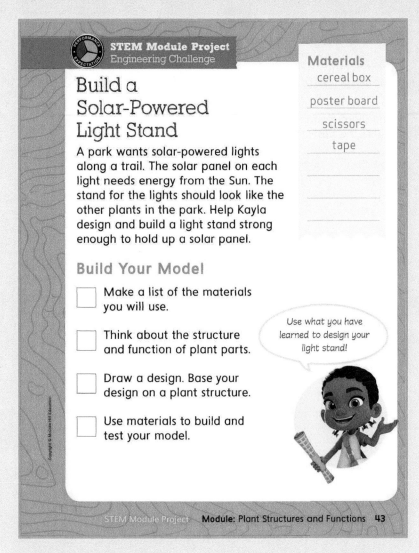

STEM Module Project
Engineering Challenge

Build a Solar-Powered Light Stand

A park wants solar-powered lights along a trail. The solar panel on each light needs energy from the Sun. The stand for the lights should look like the other plants in the park. Help Kayla design and build a light stand strong enough to hold up a solar panel.

Materials
cereal box
poster board
scissors
tape

Build Your Model

☐ Make a list of the materials you will use.

☐ Think about the structure and function of plant parts.

☐ Draw a design. Base your design on a plant structure.

☐ Use materials to build and test your model.

Use what you have learned to design your light stand!

STEM Module Project   **Module: Plant Structures and Functions**   43

🅥 GO ONLINE

INTERACTIVE PRESENTATION

Module Project: Design a Solar-Powered Light Stand

Project

ADDITIONAL RESOURCES

Module Project Rubric: Design a Solar-Powered Light Stand

☑ Rubric

 STEM Module Project
Engineering Challenge

Design a Solar-Powered Light Stand

 Prep: 15 min | **Class:** 60 min   👥 small group

Read the paragraph and explain the goal of the module project to the class. Review Lessons 1 and 2 with students. You may want to revisit the vocabulary words or phenomena from this module. Students will use engineering skills as they plan, test, and improve their design.

ASK: Why might a landscape architect want lights on a nature path? Sample answer: A landscape architect might want lights so people can see where they are walking. This will keep the plants, animals, and people safe.

ASK: Why might a park want a light that does not need to be plugged in or use batteries? Sample answer: A park might want to save energy and use the Sun to power the lights.

Project Parameters Explain to students that they will use objects like tissue boxes as substitutes for solar panels. Additionally, make sure students understand that their light stand must be at least one foot (30.48 centimeters) in the air. This will help the solar-panel capture light from the Sun. Remind students that they must design their light stand to look like a plant so it fits in with the park.

Module Project Rubric Present students with the rubric when they begin thinking about their project. Encourage students to look back at the rubric to ensure they are meeting all of the requirements of the project. Use the rubric to assess student understanding. **GO ONLINE** to access the project rubric.

Build Your Model

Students must base their light stand design on a plant structure and function.

ASK: How will your light stand be like a plant structure? Sample answer: My light stand will be tall like a plant stem.

ASK: How will your light stand function like a plant structure? Sample answer: My light stand will have roots to keep it in the ground. The stand will support the solar panel and be like a stem. My solar panels will capture sunlight like leaves.

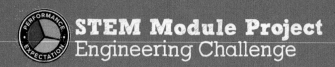

STEM Module Project
Engineering Challenge

SEP Developing and Using Models

Help students build a model light stand inspired by what they have learned about plant structures and functions. Have students start by brainstorming what types of materials they will need, drawing a design, and finally building their model.

Materials

It is recommended that you allow students to choose from a variety of different materials for this project. Here are some suggested materials: cardboard tubes, tissue box, chenille stems, scissors, tape, glue, or paper Ask parents or school janitorial staff if any of materials can be gathered from school or brought from home.

Design Your Solution

Help students create drawings of light stands. Have students label what material each portion of the light stand will be made of. Students should circle their best design and have it approved by an adult before moving forward.

Test Your Model

Have a designated area in the classroom where students can test their models to see how well the light stands hold up the solar-panels. Help students evaluate how well their light stand met the project criteria.

Have students refer to the rubric at the beginning of the project to make sure their model fits all of the criteria.

STEM Module Project
Engineering Challenge

Design Your Solution

Draw different pictures of what your light stand might look like. Circle the best one to build.

Energy

Drawing should include a sketch of a light stand with a solar panel.

Test Your Model

Explain how well your design worked.

Sample answer. I based my design on a plant stem and roots. I built a strong base and pole to hold the light up.

44 STEM Module Project   **Module: Plant Structures and Functions**

Differentiated Instruction

AL Scaffold steps to designing a solution by providing a checklist with visual supports. Possible list items: choose plant structures, draw a model, label materials, etc.

OL Have students create a list of steps for designing and building a light stand in small groups. Rotate groups in a jigsaw fashion so that students can share and gain ideas.

BL Have students think of other man-made objects that have similar design solutions to plant parts. Have them create models of these objects with left-over materials.

Module Wrap-Up

REDISCOVER THE PHENOMENON | How does this plant stay upright? Draw a picture. Add labels.

Drawing should include roots and a stem. The roots and stem should be labeled. Labels should include the function of each structure.

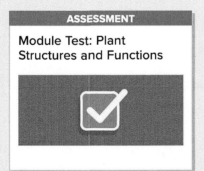

Look at your project to help you answer the question.

Module Wrap-Up   **Module:** Plant Structures and Functions   **45**

MODULE WRAP-UP
REDISCOVER THE PHENOMENON

Look back at the module phenomenon showing a young tree with its roots in the ground. Have students discuss which structures help a plant stay firmly in the ground.

Encourage students to relate this phenomenon to their module project.

ASK: How were the light stands you created similar to this plant? Sample answer: Our light stands had a strong support like a stem. The solar panels also collected sunlight like the leaves of this plant.

ASK: Why are roots important for plants that grow in very windy places? Sample answer: The roots would help hold the plant in the ground so it wasn't blown over by the wind.

GO ONLINE

INTERACTIVE PRESENTATION	ASSESSMENT
Module Wrap-Up: Plant Structures and Functions	Module Test: Plant Structures and Functions
WRAP-UP	✓

Three-Dimensional Learning

In this module, students will investigate plant survival and design a seed that can travel.

SEP Science and Engineering Practices

• Analyzing and Interpreting Data

• Constructing Explanations and Designing Solutions

DCI Disciplinary Core Ideas

• ETS1.C

• LS1.A

• LS1.D

• LS3.A

• LS3.B

CCC Crosscutting Concepts

• *Connections to Engineering, Technology, and Applications of Science*
Influence of Science, Engineering and Technology on Society and the Natural World

• Patterns

• Structure and Function

Performance Expectations

K-2-ETS1-3. **Analyze data from tests of two objects designed to solve the same problem** to compare the strengths and weaknesses of how each performs.

1-LS1-1. **Use materials to design a solution to a human problem by mimicking how plants and/or animals** use their external parts to help them survive, grow, and meet their needs.

1-LS3-1. **Make observations to construct an evidence-based account that** young plants and animals are **like, but not exactly like,** their parents.

CROSS-CURRICULAR ▸ Connections

In addition to in-depth coverage of the three dimensions, this module also covers connections to Math, English Language Arts, Writing, Engineering, and Environmental topics.

◆ GO ONLINE for Prefoessional Learning vidoes that support three-dimensional learning.

Disciplinary Core Idea Progressions

K-2	3-5	6-8
ETS1.C		
• Because there is always more than one possible solution to a problem, it is useful to compare and test designs. (K-2-ETS1-3) (secondary to 2-ESS2-1)	• Different solutions need to be tested in order to determine which of them best solves the problem, given the criteria and the constraints. (3-5-ETS1-3) (secondary to 4-PS4-3)	• Although one design may not perform the best across all tests, identifying the characteristics of the design that performed the best in each test can provide useful information for the redesign process—that is, some of the characteristics may be incorporated into the new design. (MS-ETS1-3) (secondary to MS-PS1-6)
LS1.A		
• All organisms have external parts. Different animals use their body parts in different ways to see, hear, grasp objects, protect themselves, move from place to place, and seek, find, and take in food, water and air. Plants also have different parts (roots, stems, leaves, flowers, fruits) that help them survive and grow. (1-LS1-1)	• Plants and animals have both internal and external structures that serve various functions in growth, survival, behavior, and reproduction. (4-LS1-1)	• All living things are made up of cells, which is the smallest unit that can be said to be alive. An organism may consist of one single cell (unicellular) or many different numbers and types of cells (multicellular). (MS-LS1-1)
LS1.D		
• Animals have body parts that capture and convey different kinds of information needed for growth and survival. Animals respond to these inputs with behaviors that help them survive. Plants also respond to some external inputs. (1-LS1-1)		
LS3.A		
• Young animals are very much, but not exactly like, their parents. Plants also are very much, but not exactly, like their parents. (1-LS3-1)	• Many characteristics of organisms are inherited from their parents. (3-LS3-1)	• Genes are located in the chromosomes of cells, with each chromosome pair containing two variants of each of many distinct genes. Each distinct gene chiefly controls the production of specific proteins, which in turn affects the traits of the individual. Changes (mutations) to genes can result in changes to proteins, which can affect the structures and functions of the organism and thereby change traits. (MS-LS3-1)
LS3.B		
• Individuals of the same kind of plant or animal are recognizable as similar but can also vary in many ways. (1-LS3-1)	• Different organisms vary in how they look and function because they have different inherited information. (3-LS3-1)	• In addition to variations that arise from sexual reproduction, genetic information can be altered because of mutations. Though rare, mutations may result in changes to the structure and function of proteins. Some changes are beneficial, others harmful, and some neutral to the organism. (MS-LS3-1)

Three Dimensions at a Glance

Throughout this module and in the culminating module project, students will integrate relevant Science and Engineering Practices and Crosscutting Concepts into their learning and understanding of the Disciplinary Core Ideas. Use this chart to locate where students will encounter each of the three dimensions that build to the Performance Expectations.

DIMENSIONS	LESSON 1	LESSON 2	MODULE PROJECT
SEP Analyzing and Interpreting Data (K-2-ETS1-3)	•		•
SEP Constructing Explanations and Designing Solutions (1-LS1-1), (1-LS3-1)	•	•	
DCI **ETS1.C** Optimizing the Design Solutions (K-2-ETS1-3)			•
DCI **LS1.A** Structure and Function (1-LS3-1)		•	
DCI **LS1.D** Information Processing (1-LS3-1)		•	
DCI **LS3.A** Inheritance of Traits (1-LS1-1)	•		
DCI **LS3.B** Variation of Traits (1-LS1-1)	•		
CCC *Connections to Engineering, Technology, and Applications of Science* Influence of Science, Engineering and Technology on Society and the Natural World (1-LS1-1)		•	
CCC Patterns (1-LS3-1)	•		
CCC Structure and Function (1-LS1-1)		•	

Module Planner

In this module, students will compare plants across generations and learn how plants survive.

	Module Opener	Lesson 1: Plants and Their Parents	Lesson 2: Plant Survival
	Big Idea: How do plants grow and survive?	**Essential Question:** Are plants and their parents the same?	**Essential Question:** How do plants survive?
Pacing 1 Day = 45 min	0.5 Day	7 Days	7 Days
Summary	In this module, students will develop the understanding that plants have almost identical structures to their parents and that these structures help the plants survive.	Students will observe similarities and differences between plants and their offspring.	Students will learn how plant structures help plants survive.
Inquiry Activity		**Data Analysis** Compare an Adult Plant and a Young Plant **Simulation** Plants Grow and Change **Hands On** Grow a Radish	**Simulation** Plants and Shade **Research** Plant Survival
Vocabulary		inherit, offspring, parent, seedling	need, pollen, survive
Cross-Curricular Connections		ELA, MATH	ELA, Engineering, Math

School-to-Home Resources

🕹 **GO ONLINE** for the following resources to strengthen the school-to-home connections.

Letter to Home will help parents and guardians understand the learning objectives for the Plant Parents and Their Offspring module.

STEM Module Project: Design a Seed That Travels	Module Wrap-Up
2 Days	.05 Day
Students will use what they've learned throughout the module to design a seed that can travel. They will build and test a model seed.	Students will revisit the Module phenomenon and explain their learning.
Math, Engineering	

Assessment Tools

Formative Assessment
Includes Page Keeley Science Probes; Claim, Evidence, Reasoning; Three-Dimensional learning checks

STEM Module Project
Authentic performance-based assessment with rubric

McGraw-Hill Assessment
Ready-made assessments that can be printed or delivered electronically

Inquiry Activity Planner

In this module, students will compare plant structures in plants and their offspring, then determine how plants stay alive.

Lesson	Inquiry Activity		Materials	
	★ ⬤ GO ONLINE for teacher support videos on selected activities. Materials included in the Collaboration Kit are listed in blue.		**Consumable**	**Non-Consumable**
Lesson 1	**Data Analysis** Compare an Adult Plant and a Young Plant	🕐 30 min 👥 pairs		
	Purpose: Students will compare a young oak and an adult oak.			
	Simulation Plants Grow and Change	🕐 30 min 👥 small groups		
	Purpose: Students will investigate how several species of plants change as they grow. **Plan Ahead:** Reserve technology cart, if needed.			
	Hands On Grow a Radish	🕐 30 min 👥 small groups	plastic cup, potting soil, radish seeds, water, plastic gloves	cubes
	Purpose: Students will observe how a radish changes as it grows and develops.			
Lesson 2	**Simulation** Plants and Shade	🕐 30 min 👥 small groups		
	Purpose: Students will investigate whether all plants need the same amount of sunlight. **Plan Ahead:** Reserve technology cart, if needed.			
	Research Plant Survival	🕐 35 min 👥 small groups		
	Purpose: Students will research mechanisms plants have that help them survive.			
STEM Module Project	**Engineering Challenge** Design a Seed That Travels	🕐 60 min 👥 small groups	Suggested: chenille stems, cotton balls, hook and loop fastener, polystyrene foam, tissue paper, tape, toothpicks construction paper, glue	

McGraw-Hill Education is your partner for hands-on materials! To order new Collaboration Kits or refill specific items, contact the McGraw-Hill customer service line at (800) 338–3987.

Inquiry Activity Support

Guides activities with confidence by watching the Inquiry Activity Preview video as you plan the day's Inquiry Activity. After your students complete the activity, give them all a common set of expected observations by showing them the Inquiry Rewind video.

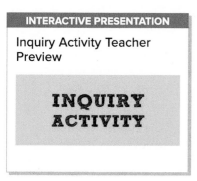

INTERACTIVE PRESENTATION

Inquiry Activity Teacher Preview

INQUIRY ACTIVITY

As you plan each Explore Inquiry Activity, watch this video for information about Activity setup, strategies for a smooth activity experience, and math and skill reviews you can do to launch the activity and prepare your students to complete the activity successfully.

INTERACTIVE PRESENTATION

Inquiry Activity Rewind

INQUIRY REWIND

Each Inquiry Activity Rewind video shows students a step-by-step procedure and expected or sample results. Highlight important observations students might have missed during the activity so, even if they missed class, everyone is ready to interpret the data and construct explanations

Teacher Notes

Inspire All Students

Strategies to scaffold your instruction and plan for successful teaching for all students.

Differentiated Instruction

Module Concept Plants and animals and their offspring are alike, but not exactly alike, and both have certain behaviors that help offspring survive. Help students to connect these concepts by providing multiple means of engagement.

AL Approaching Level

Bring in a selection of fruits and vegetables. Allow students to examine them and discover which ones have seeds and which do not. Have each student choose a fruit or vegetable and draw a picture of the plant they think it came from. Ask them to name the plant and explain how they know.

OL On Level

Have students make a chart to categorize the examples as fruits and vegetables. Then share with students that a fruit is what carries the seeds of a plant. Have students adjust their charts to reflect this new information. Where will they put green beans on the chart? Have students explain their thinking.

BL Beyond Level

Have students predict how spinach or another leafy green vegetable reproduces. Is it by a seed? If so, where does the seed reside? If not, how are new plants made? Have students draw pictures to accompany their predictions. Allow them to research the answers to confirm or disprove their predictions. Then have them report to the group.

Advanced Learners and Gifted Learners

Instruction should focus on adding depth and complexity in understanding how plant offspring are different from their parents and investigating how the structure and function of young plants helps them grow and survive.

DOK 3 Strategic Thinking Have students think back to one of the Inquiry Activities in the module and create a clay model of the plant that was featured in that activity. Then have them use vocabulary from the lesson to describe or label their models. For example, for the Plant Survival Inquiry Activity from Lesson 2, students could create a clay model of a rose and explain how to roots help the rose meet its needs or how thorns help it survive

DOK 4 Extended Thinking Have students analyze the structures of plants that thrive in different climates. How are the parts similar or different? For example, are roots shallow or deep? Does the plant have leaves or needles? Flowers? Alternatively, have students examine an unknown plant and determine what climate it might come from, based on its structures. Have them explain to a partner why they think so, using module vocabulary if relevant.

Literacy Support: Using the Leveled Readers

Use the Leveled Readers to enable students to further develop their literacy skills through science.

- Fiction: Engages students in key concepts.
- Nonfiction: Focuses on real-world topics; Makes informational text accessible to all learners.
- Also available in print and online.

How Plants Survive

Summary This books describes how plants grow and survive.

When to Use Use during Lesson 2 to further develop an understanding of how plants survive.

Martin Wahlborg/E+/Getty Images

English Language Support

EMERGING

Brainstorm Web After writing a word or phrase in a circle on chart paper, have students write as many words that they already know connected to this word or phrase as they can think of. Allow students to brainstorm orally first, using home language as needed, then have them write the words. For example, write Plants and their offspring in the center and have students write words they know related to plants and their offspring, for example, leaf, stem, seed. Keep the web visible throughout the module and have students add to it as you move through the lessons.

EXPANDING

Anchor Chart Have students activate prior knowledge by writing sentences to place on an Anchor Chart. Provide sentence starters that require students to write what they know about plants, for example: One thing I already know about plants and their offspring is [they usually look alike]. or One important thing I think about plants and their offspring is [they both need sunlight]. As you move through the module, have students add sentences to the chart displaying what they've learned.

BRIDGING

K-W-L Chart In pairs, have students discuss what they already know about plants and their offspring, for example: Plants and their offspring look similar but their offspring are small. They both need air, sun, and water to survive. Then, as a class, have students tell you what they "know" and list it in the K section of the chart. Then discuss what the students "want" to know, for example, how parents and their offspring are different, and write the questions in the W section of the chart. As you go through the lesson, add what students have "learned" in the L part of the chart.

Cognates

Cognates are words in two different languages that share a similar meaning, spelling, and pronunciation. Review differences in spelling and pronunciation of these terms with your Spanish-speaking English learners.

inherit *heredar*	**parent** *padre*	**survive** *sobrevivir*
pollen *polen*	**need** *necesitar*	**adult** *adulto*
bulb *bulbo*	**soil** *suelo*	

Online Vocabulary Resources

The Vocabulary Organizers are intended to support English language learning and vocabulary acquisition. Resources for each stage of the learning process are meant to appeal to different types of learning styles (kinesthetic, tactile, auditory, written, visual) and provide multiple modalities of repeated exposure.

Explore	Study	Review
Vocabulary Concept Circle	Frayer Model	Science Vocabulary

Performance Expectations

The learning experiences throughout the module will develop student understanding of the following Performance Expectations:

1-LS1-1. Use materials to design a solution to a human problem by mimicking how plants and/or animals use their external parts to help them survive, grow, and meet their needs.

1-LS3-1. Make observations to construct an evidence-based account that young plants and animals **are like, but not exactly like,** their parents.

K-2-ETS1-3. Analyze data from tests of two objects designed to solve the same problem to compare the strengths and weaknesses of how each performs.

Plant Parents and Their Offspring

STEM Connections

⚫ **GO ONLINE** to see STEM Connections, a diverse selection of people and groups that have made important contributions to society through science and technology.

Teacher Toolbox

Science Background

This photo shows a dandelion being blown by the wind. The seeds on the dandelion have structures called parachutes that help them float in the wind. Students should return to this photo at the end of the module when they design their own seed that moves.

Identifying Preconceptions

Students may think that plants behave the same way as humans. Due to the abundance of anthropomorphic stories, students may think that plants have human characteristics. However, plants do not think in the same way humans do. They cannot choose to have offspring or move to a new location. They receive signals from their environment and respond automatically.

DISCOVER
THE PHENOMENON

What happens when you blow on a dandelion?

Dandelion Seeds **GO ONLINE**
Watch the video *Dandelion Seeds* to see the phenomenon in action.

🗨 Talk About It

Look at the photo.

Watch the video.

What do you observe?

What questions do you have?

Did You Know?

The part of the dandelion that floats away when you blow it is the seed!

Module: Plant Parents and Their Offspring **47**

🅖 GO ONLINE

INTERACTIVE PRESENTATION

Discover the Phenomenon: Plant Parents and Their Offspring

DISCOVER THE PHENOMENON

 10 min   whole class

Science often begins when someone makes an observation about a situation or occurrence. Scientists refer to an event or situation that is observed or can be studied as a phenomenon. Have students study the picture of the dandelion blowing in the wind.

Ask the **Discover the Phenomenon** question:

What happens when you blow on a dandelion?

This leads to the overarching module **Big Idea:**
How do plants grow and survive?

🅖 **GO ONLINE** Check out *Dandelion Seeds* to see this phenomenon in action.

🗨 Talk About It

Ask students to describe what they see. Help students turn their observations from the video into questions. Record observations on the board or chart paper. Start a class discussion with the following prompts:

• What object does a dandelion's seed remind you of?

• Why do you think dandelion seeds blow away in the wind?

• Why don't the seeds fall directly onto the ground?

Record responses and questions on the board or chart paper to look at as you move through the module.

Did You Know?

After a dandelion has flowered, a seed head develops that usually contains over a hundred seeds.

MODULE OPENER

STEM CAREER Connection

What Does a Farmer Do?

 15 min whole class

Introduce the farmer STEM CAREER Connection. Encourage students to look at the photos and share their observations. Throughout this module, students should apply the skills of a farmer to their learning. Remind students that farmers study generations of plants and animals.

PRIMARY SOURCE

George Washington Carver was a pioneer of sustainable farming and environmentalism. Carver was born into slavery in 1864 and rose to become an important botanist, inventor, and teacher. After realizing that soil was severely depleted by the repeated plantings of cotton, Carver popularized the concept of crop rotation. He encouraged farmers to plant peanuts and soybeans to help restore the nitrogen in the soil between cotton plantings.

STEM CAREER Connection

What Does a Farmer Do?

Farmers grow plants and raise animals. Some farmers grow crops. Crops are plants that are grown and harvested. Farmers can grow grains, fruits, or vegetables.

Some farmers raise livestock. Livestock are animals raised for human use. Animals like sheep, cows, and chickens are livestock.

PRIMARY SOURCE

George Washington Carver was a botanist, inventor, and professor. He taught farmers how to grow crops that would keep the soil healthy.

48 STEM CAREER Connection **Module:** Plant Parents and Their Offspring

🖰 GO ONLINE

INTERACTIVE PRESENTATION

STEM Career: Farmer

STEM CAREER

Differentiated Instruction

AL Help students brainstorm a list of things a farmer might do every day: How does a farmer grow his crops and keep them healthy?

OL Ask students what other careers involve working with plants. Have them discuss why knowledge of plants would be important in that career.

BL Have partners research hydroponics and study photos of indoor farms. Ask: How is this type of farming the same or different from traditional farming?

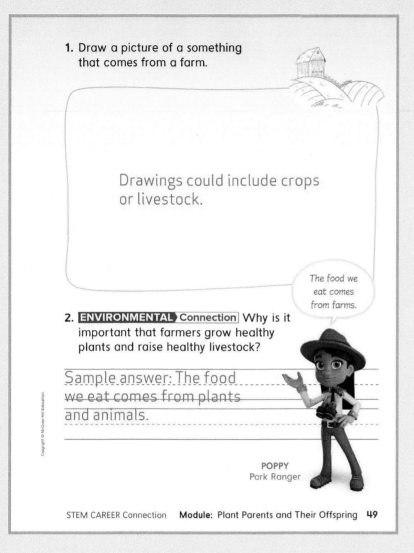

1. Draw a picture of a something that comes from a farm.

Drawings could include crops or livestock.

The food we eat comes from farms.

2. **ENVIRONMENTAL** Connection Why is it important that farmers grow healthy plants and raise healthy livestock?

Sample answer: The food we eat comes from plants and animals.

POPPY
Park Ranger

STEM CAREER Connection   **Module:** Plant Parents and Their Offspring   **49**

Have students answer question 1 and share their drawings with classmates. Encourage students to explain what they drew and how it relates to a farm.

Discuss with students that nature makes and does things that help humans. Have students think about why humans need healthy plants in the environment.

ASK: What are some things we get from nature? Sample answer: Fruits and vegetables.

ASK: What are some things we make using things from nature? Sample answer: We make paper from the trunks of trees. We also make clothing from plants like cotton.

ASK: What are some things nature does for us? Sample answer: Trees and other plants make oxygen for us so we can breathe.

Have students read Poppy's speech bubble.

Word Wall

 10 min whole class

The words below represent some of the fundamental words from this module. By the end of this module, students should be able to use these words correctly in context.

Write each vocabulary word on the board. As you write each word, have students pronounce the word. Have students describe the photo next to each word to get them thinking about what the word might mean. Record student responses on the board next to the word. Revisit the terms throughout the module as the students learn their meaning in context with the module content.

STEM Vocabulary

As students encounter each word, have them use context clues to determine their meaning. Use the information below to help students develop a clear understanding of each word and build their academic vocabulary.

observe In first grade, students should be able to make, record, and share observations. Ask students to play a game with a partner using observations. One person should choose an object in the room and describe it without naming the object. Encourage the partner to try to guess the object their partner describes.

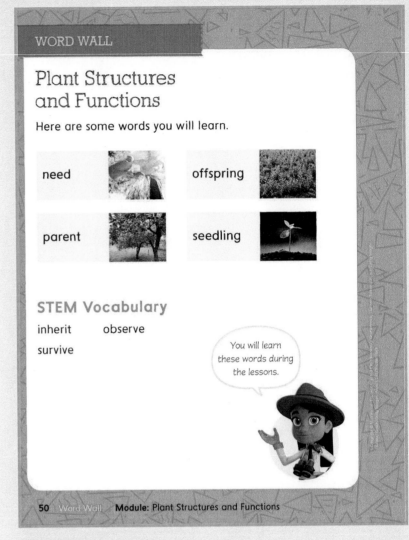

WORD WALL

Plant Structures and Functions

Here are some words you will learn.

need

offspring

parent

seedling

STEM Vocabulary

inherit   observe

survive

You will learn these words during the lessons.

50 Word Wall   **Module: Plant Structures and Functions**

🅫 GO ONLINE

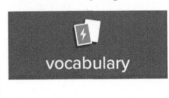

INTERACTIVE PRESENTATION

Word Wall: Plant Parents and Their Offspring

vocabulary

ASSESS PRIOR KNOWLEDGE

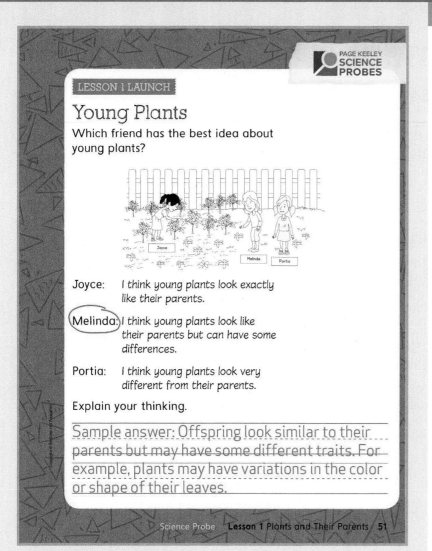

Young Plants

Which friend has the best idea about young plants?

Joyce: I think young plants look exactly like their parents.

Melinda: I think young plants look like their parents but can have some differences.

Portia: I think young plants look very different from their parents.

Explain your thinking.

Sample answer: Offspring look similar to their parents but may have some different traits. For example, plants may have variations in the color or shape of their leaves.

Science Probe   **Lesson 1 Plants and Their Parents / 51**

GO ONLINE

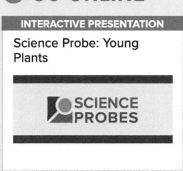

INTERACTIVE PRESENTATION

Science Probe: Young Plants

PAGE KEELEY
SCIENCE PROBES

Young Plants

 10 min   Whole class

Using the Probe

The purpose of this probe is to identify students' prior knowledge about plants and their parents. This probe works well with a four-corner strategy. Have students think about and mark their answers in their notebook. Label three corners of the room with different answer choices. Have students move to the corner they agree with. Give students time to speak with the people in their corner and construct an explanation of why they think they are correct. Have each group share their ideas. Use this probe to assess students' prior knowledge of the lesson content and to identify possible misconceptions.

Be sure not to tell students the answer. It is not important that students know the answer to this probe at this point in the lesson. What is important is the reasoning they provide to support their answer. Students will revisit the probe throughout the lesson to see how their thinking has changed.

Throughout the Lesson

Use students' explanations to bridge their initial ideas about parent plants and offspring with the understanding the students develop during the lesson. Prompts in the Teacher's Edition will instruct you when it's time for students to revisit this probe.

Teacher Explanation

The best answer is Melinda: I think young plants look like their parents but can have some differences. Offspring of sexually reproduced organisms look similar to their parents but have some differences in traits. For example, some plants may have variations in color or the shape of their leaves.

Teacher Toolbox

Identifying Preconceptions

This probe is designed to reveal whether students recognize the similarities and differences between adult plants and their offspring. Students may not realize that plants look similar to their parents, but are not exactly the same. Variations occur due to the mixing of genetic traits and random mutations.

Building to the Performance Expectations:

In this lesson, students will explore content and develop skills leading to mastery of the following Performance Expectation:

1-LS3-1. Make observations to construct an evidence-based account that young plants and animals are like, but not exactly like, their parents.

SEP Science and Engineering Practices

Construct Explanations and Designing Solutions

Constructing explanations and designing solutions in K–2 builds on prior experiences and progresses to the use of evidence and ideas in constructing evidence-based accounts of natural phenomena and designing solutions. (1-LS3-1)

DCI Disciplinary Core Idea

LS3.A: Inheritance of Traits

Young animals are very much, but not exactly like, their parents. Plants also are very much, but not exactly, like their parents. (1-LS3-1)

LS3.B: Variation of Traits

Individuals of the same kind of plant or animal are recognizable as similar but can also vary in many ways. (1-LS3-1)

CCC Crosscutting Concept

Patterns

Patterns in the natural and human designed world can be observed, used to describe phenomena, and used as evidence. (1-LS3-1)

ELA/Literacy Connections	Math Connections
W.1.8 Writing	1.MD.A.2 Measurement and Data

* See correlation table for full text of ELA and Math standards.

Track Your Progress to the Performance Expectations

You may want to return after completing the lesson to note concepts that will need additional review before your students start the module Performance Project.

Dimension	Concepts to Review Before Assessment
SEP Constructing Explanations and Designing Solutions (1-LS3-1)	
DCI LS3.A: Inheritance of Traits (1-LS3-1)	
DCI LS3.B: Variation of Traits (1-LS3-1)	
CCC Patterns (1-LS3-1)	

Lesson at a Glance

Full Track is the recommended path for the complete lesson experience. FlexTrack A and FlexTrack B provide timesaving strategies and alternatives.

	Day-to-Day	Pacing	Full Track 45 min/day (full year) Resources
Assess Prior Knowledge	Page Keeley Science Probes: *Young Plants*	Day 1	Page 51
Engage	Discover the Phenomenon: Are these plants the same?		Pages 52–53 Video: *A Growing Plant*
Explore	Inquiry Activity: *Compare an Adult Plant and a Young Plant*	Day 2	Pages 54–56
	Science Read Aloud: *Perfect Acorn, Mighty Oak*		Page 57
Explain	Seedlings Grow Science Read Aloud: *Every Plant is Different*	Day 3	Pages 58–59 Video: *Plant Parents and Their Offspring* Go Further activity: *Plants and Their Offspring Are Alike and Different*
	Close Reading: *Young Plants Are Like Their Parents*	Day 4	Pages 60–61
	Inquiry Activity: *Plants Grow and Change*	Day 5	Pages 62–63
Elaborate	Inquiry Activity: *Grow a Radish*	Day 6	Pages 64–65
Evaluate	Explain the Phenomenon: Are these plants the same?	Day 7	Pages 66–68
			7 Days

Essential Question: Are plants and their parents the same?

Objective: Make observations and explain that young plants and their parents are similar but not exactly alike.

Vocabulary: inherit, offspring, parent, seedling

FlexTrack A	
30 min/day (5 days per week)	
Pacing	**Resources**
Day 1	Page 51 Employ the Fish Bowl strategy.
	Pages 52–53 Video: *A Growing Plant*
Day 2	Pages 54–56
Day 3	Page 57
Day 4	Pages 58–59 Video: *Plant Parents and Their Offspring*
Day 5	Pages 60–61 Omit the VKV.
Day 6	Pages 62–63 Use the simulation as a teacher demonstration.
Day 7	Page 66–68 Answer the Explain the Phenomenon question as a class.
7 Days	

FlexTrack B	
30 min/day (3 days per week)	
Pacing	**Resources**
Day 1	Page 51 Employ the Fish Bowl strategy.
	Pages 52–53 Video: *A Growing Plant*
Day 2	Pages 54–56
Day 3	Pages 60–61 Omit the VKV.
Day 4	Pages 66–68 Answer the Explain the Phenomenon question as a class.
4 Days	

Lesson 1: **Plants and Their Parents**

Lesson Objective

Students will explore plants and their offspring to identify similarities and differences. They will construct explanations about patterns in the natural world.

DCI Inheritance of Traits

LS3.A Young animals are very much, but not exactly like, their parents. Plants are also very much, but not exactly, like their parents.

DCI Variation of Traits

LS3.B Individuals of the same kind of plant or animal are recognizable as similar but can also vary in many ways.

LESSON 1

Plants and Their Parents

52 Module: Plant Parents and Their Offspring

Teacher Toolbox

Science Background

To the untrained eye, plants and their offspring often look very similar. While some changes, such as color variation, are easy to notice, other types of variation only become apparent while examining a plant's DNA. Sexually reproducing plants inherit a combination of traits and characteristics from their parent plants. This leads to variation in the offspring. Likewise, even asexually reproducing plants can have spontaneous mutations that result in new characteristics from one generation to the next. Have students examine the photo and look for similarities and differences between the plants they observe.

DISCOVER
THE PHENOMENON

Are these plants the same?

 A Growing Plant

GO ONLINE

Watch the video *A Growing Plant* to see the phenomenon in action.

Look at the photo. Watch the video. How are young plants and adult plants similar? What do you observe?

Sample answer: I think these plants are the same type of plant but different ages. Young plants and adult plants have many of the same structures. They both have stems and roots.

Did You Know?
Almost all plants grow from tiny seeds!

Engage   **Lesson 1** Plants and Their Parents   **53**

GO ONLINE

INTERACTIVE PRESENTATION

Discover the Phenomenon: Plants and Their Parents

DISCOVER THE PHENOMENON

 10 min   whole class

Recall that scientists refer to an event or situation that is observed or can be studied as a phenomenon. Have students study the photo of the greenhouse to make observations about the plants they see.

Ask the **Discover the Phenomenon** question:
Are these plants the same?

➡ This leads to the overarching lesson **Essential Question:**
How are plants similar and different from their offspring?

GO ONLINE Check out *A Growing Plant* to see the phenomenon in action.

Talk About It

Ask students to describe what they see. Help students turn their observations from the video into questions. Start a class discussion with the following prompts:

ASK: How are the plants similar and different?

ASK: How do the plants change as they grow?

Record responses and questions on the board or chart paper to refer to as you move through the module.

Did You Know?

Mosses and ferns grow from spores rather than seeds. Potato plants can also grow with or without seeds. Potatoes grown by planting a potato in the soil will be genetically identical to their parent plant, while potatoes grown from seeds will be unique.

Lesson 1: **Plants and Their Parents**

INQUIRY ACTIVITY | Data Analysis

Compare an Adult Plant and a Young Plant

 Prep: 2 min I **Class:** 30 min   pairs

Purpose
Students will compare a young oak and an adult oak.

Materials
Alternative: If an adult plant and its offspring are available, these can be substituted in this activity.

Before You Begin
Remind students of the greenhouse plants they observed in the lesson phenomenon. Have students think about young plants and their parents. Focus students on the question they will investigate:

ASK: How are a young and an adult oak tree alike and different?

Post the question to revisit as a class during the activity.

Guide the Activity
Make a Claim Students will use their prior knowledge to write a claim describing the similarities and differences between a young oak and an adult oak tree.

Investigate
1. Read the photo captions with students. Make sure they understand whether the photos show a young oak or an adult oak.

Have students complete the Compare and Contrast Graphic Organizer from page EM18 as they observe the adult and young plant.

INQUIRY ACTIVITY

Data Analysis

Compare an Adult Plant and a Young Plant

Plants change as they grow. Use the photos to observe how an oak tree changes.

Make a Claim How are a young and an adult oak tree alike and different?

Sample answer: The young oak and adult oak have the same leaf shape. An adult oak has acorns and a young oak does not.

Investigate

1. Look at the photos of the young oak tree and the adult oak trees.

2. **Record Data** Circle the parts that are the same.

3. Place an X on the parts that are different.

This is a young oak tree. It grew from a seed.

GO ONLINE

INTERACTIVE PRESENTATION

Inquiry Activity: Compare an Adult Plant and a Young Plant

INQUIRY ACTIVITY

ADDITIONAL RESOURCE

Inquiry Rewind: Compare an Adult Plant and a Young Plant

INQUIRY REWIND

This is an adult oak tree.

Acorns grow on adult oak trees. The acorn is the fruit of the adult plant. There is a seed inside the acorn.

Strong wind blew this oak tree out of the ground. You can see the roots of the adult oak tree.

GO ONLINE

ADDITIONAL RESOURCE

Science Song: Patterns

CCC Patterns

Patterns exist everywhere. In first grade, students should be classifying objects. This can be done by looking for similarities and differences between plants. Students should begin using patterns to predict relationships between organisms.

ASK: What made you think these two plants were related? Sample answer: These two plants had very similar parts. The shape of leaves made me think they were related.

▶ **GO ONLINE** Have students listen to the Crosscutting Concepts Science Song: *Patterns*.

Crosscutting Concepts Graphic Organizer

▶ **GO ONLINE** Use the Crosscutting Concept Graphic Organizer to identify patterns between young and adult plants.

SEP Analyzing and Interpreting Data

Students in first grade should be collecting, recording, and sharing observations. This will help them with their ability to plan and carry out investigations. Encourage students to determine whether their observations support the claims they wrote.

ASK: What similarities did you see between the young plant and the parent plant? Sample answer: I saw that both the young oak and adult oak have leaves and trunks.

ASK: What differences did you see between the young plant and the parent plant? Sample answer: The young plant had a thinner trunk than the adult plant.

Inquiry Spectrum

Structured Inquiry

Guide students through this activity as it is written in the student notebook. Have students compare images of a young oak and an adult oak.

Guided Inquiry

Ask students to cover up the steps of the investigation. Present students with the objective and the claim. Ask students to develop a procedure to collect evidence to support or refute their claim.

Open Inquiry

Show students an image of a field filled with plants of different ages. Ask students what questions they have about the plants. Have students design an investigation to answer their question.

INQUIRY ACTIVITY | Data Analysis

Compare an Adult Plant and a Young Plant *(continued)*

Communicate

4. Help students brainstorm why an adult oak tree with small roots might become problematic. Remind students of the vocabulary words *structure* and *function*.

5. There are many possible outcomes for this activity. Students may find that their observations supported their claim, refuted their claim, or that there is not enough information to either support or refute their claim. If the latter is the case, begin a discussion about what further research could be conducted to determine the validity of the claim.

Have students discuss how the information they've learned relates to other plants.

 Talk About It

Have students brainstorm ideas as a class. Encourage students to explain their thoughts using details from this investigation and background knowledge.

Short on Time?

Do the activity as a whole class. Read and discuss the photo captions as a class. Give students time to discuss their answer to the questions and record a set of class responses together.

Communicate

4. How are the roots of the young and adult oak trees different?

Sample answer: The roots of the adult oak tree are much thicker and longer.

5. What do you think would happen if the roots of the adult tree stayed the same size as the roots of the young tree?

Sample answer: The roots would not be able to hold the adult tree in place.

6. Did your data match your claim? Explain.

Sample answer: Yes. The young oak and adult oak trees have the same leaf shape. An adult oak tree has acorns and a young oak tree does not.

 Talk About It

How else do you think a young plant and adult plant are similar and different? Tell a partner.

Differentiated Instruction

AL Have students draw the stages of the life cycle of the oak or other tree, from seed to seedling, to adult tree. When they are done, help them use lesson vocabulary to describe what is happening at each stage of the process.

OL Have partners identify similarities and differences about the oak at various stages of its life on pages 54 and 55 and make compare-and-contrast statements using signal words, such as *and, both, the same as* or *but, however, is different from.* For example, An adult tree has fruit, but a seedling does not. Both and adult and a seedling have leaves.

BL Have partners discuss the purpose of an acorn shell and how it is like a bicycle helmet. Then have them think of other examples of things that have a protective shell.

Listen to *Perfect Acorn, Mighty Oak.*

Oak trees provide food and shelter to animals.

7. Why does Gary plant an acorn and not a leaf?

Sample answer: New plants grow from seeds. There are seeds inside acorns. An oak leaf will not grow into a new plant.

8. How does your inquiry activity match what you read?

Sample answer: As the tree grows the trunk becomes thicker, the tree grows taller, and there are more leaves.

Explore   **Lesson 1 Plant and Their Parents   57**

🐦 GO ONLINE

INTERACTIVE PRESENTATION

Read Aloud: Perfect Acorn, Mighty Oak

Compare an Adult Plant and a Young Plant *(continued)*

🕑 20 min   👥 whole class

Before Reading

Have students observe the picture on page 4 of the Science Read Aloud *Perfect Acorn, Mighty Oak.* Ask students to share how the picture is similar to the oak trees they saw in their student notebook.

💬📖 Science Read Aloud

Read aloud with students pages 4–13 of the Science Read Aloud *Perfect Acorn, Mighty Oak.* Students will learn how an oak tree changes as it ages.

After Reading

Ask students to summarize what happened in the reading. Point out that students have already compared a young oak tree to an adult oak tree.

WRITING › Connection

W.1.8 Research to Build and Present Knowledge
Prompt students to think about their past experiences with plants. Encourage students to use their Science Read Aloud as well as their own background knowledge about plants to answer question 7.

EL Support

ELD.PI.1.1, SEP-1 Preview the text before reading the selected Read Aloud. Introduce/review the terms: *leaves, bark, branch, acorn, stem.* Pair students with an English-proficient student for collaborative conversation. Have partner ask questions at an appropriate level for the English learner.

EMERGING
Have English-proficient students ask questions to which students answer by pointing or saying *yes* or *no.* Where is an acorn? Is this a branch?

EXPANDING
Have English-proficient students ask questions requiring a simple sentence as an answer. What is this? It is an acorn.

BRIDGING
Have English-proficient students ask more complex questions. Where are the acorns? The acorns are in the tree and on the ground.

Lesson 1: **Plants and Their Parents**

Seedlings Grow

 30 min   whole class

VOCABULARY

Encourage students to use context clues to derive the meaning of vocabulary words. Struggling students can use the glossary to access the definitions.

GO ONLINE to watch the video *Plant Parents and Their Offspring* individually or play the video for the whole class.

Scientific Vocabulary

seedling Tell students the suffix -ling is often used to show that a person or thing has a specific origin, or that it comes from something else. Therefore, the plant that grows from a seed, is referred to as a seedling. Duckling and nestling are other nouns that use the suffix -ling in this way.

Academic Vocabulary

change Students may be familiar with the word *change* as it is used as a noun. Students may consider *change* to mean the money given back when a purchase is made using an amount of money larger than the cost of the object. The word *change* as a noun means transformation. When one thing is traded for another, we call this an exchange. Ask students to discuss how the word exchange is similar to the word *change*.

Teacher Toolbox

Science Background

Despite the various ways that plants reproduce, similarities exist between adult plants and their offspring. Plants of the same species will share more genetic traits than plants of another species. The differences that occur from one generation to the next can be due to the mixing of parental genes, random mutations, or changes caused by the environment where they grow. The degree of genetic similarity can be used to tell how closely related plants are.

Identifying Preconceptions

Students may believe that young plants are an entirely different species than the parent plant. They may be unfamiliar with plant development and fail to recognize that plants change as they grow. Likewise, students may not understand that all plants are related and share some similarities. They may not realize that the more similar plants appear, the more likely it is that they are closely related.

---

58   Module: **Plant Parents and Their Offspring**

Vocabulary

Look and listen for these words as you learn about plants and their parents:

inherit   offspring   parent   seedling

Seedlings Grow

A **seedling** is a young plant. Seedlings grow into adult plants.

GO ONLINE

Watch the video *Plant Parents and Their Offspring* to learn about plants and their parents.

1. How does this plant change as it grows?

Sample answer: The stems grow stronger. The leaves become larger. The adult plant has flowers.

58   Explain   **Module:** Plant Parents and Their Offspring

GO ONLINE

INTERACTIVE PRESENTATION	ADDITIONAL RESOURCE
Read About: Seedlings Grow	**Lesson Vocabulary: Plants and Their Parents**

vocabulary

Listen to *Every Plant is Different*.

Young plants are similar and different from their parent plant. You can observe these differences. Plants change as they grow.

2. How might a young plant be different from its parent? Make a list.

Sample answer: color | size of trunk

number of leaves | height

3. What do you think this sunflower looked like as a seedling? Draw a picture.

Drawing should include a stem, leaves, and roots. The stem of the seedling should be thin, there should be fewer leaves, and no flower.

◆ GO ONLINE

Learn more by exploring *Plants and Their Offspring Are Alike and Different*.

Explain   **Lesson 1** Plants and Their Parents **59**

◆ GO ONLINE

INTERACTIVE PRESENTATION

Read Aloud: Every Plant is Different

Before Reading

Have students observe the photo on page 14 of the Science Read Aloud *Every Plant is Different*. Remind students that young and old plants have similarities and differences. Ask them to brainstorm lists of how young and old plants are similar and different.

Science Read Aloud

Read aloud with students pages 14–24 of the Science Read Aloud. Students will learn about plants and their parents. While reading, students will encounter the vocabulary words: *inherit, parent,* and *seedling*.

After Reading

Have students compare the fiction and nonfiction texts. Ask how these two texts were similar. Have students discuss how adult plants and their seedlings are alike and different. Encourage students to share how these readings relate to their own life and experiences.

2. Help students create a list of ways a young plant is different from its parent.

Visual Literacy

Read a Diagram Have students study the photo on page 59. Tell students to draw a picture of what they think the sunflower looked like when it was young.

◆ **GO ONLINE** and use *Plants and Their Offspring Are Alike and Different* to learn more.

Differentiated Instruction

AL Help students make a list of five fruits and identify where the seeds can be found for each. For example, An apple's seeds are found inside the core.

OL Have partners explain why seeds are usually found inside a fruit and not elsewhere. What does the fruit provide to the seed that helps the plant reproduce?

BL Have students make other true/false statements about plants and their parents. The partner should say whether the statement is true or false. If it is false, they should amend the statement so that it is true.

CLOSE READING

Offspring Are Like Their Parents

 30 mins   partners

Inspect

Read Have students read the passage to focus on understanding the overall meaning. Tell students to take notes to gather information about plants and their offspring. Tell students to think about how young plants and their parents are different.

Find Evidence

Have students reread the text. Tell them to underline differences between plant parents and young plants, and to circle similarities.

VKV Visual Kinesthetic Vocabulary

 15 min   whole class

Have students cut out and fill in the Dinah Zike Visual Vocabulary from pages VKV3-VKV4 in their notebook. Guide students in using the VKV to learn the words *offspring*, *parent*, and *inherit*.

CLOSE READING

Inspect

Read *Offspring Are Like Their Parents.* Think about the differences between offspring and their parents.

Find Evidence

Underline how offspring and their parents are different. Circle how they are the same.

Notes

_____

_____

_____

_____

Offspring Are Like Their Parents

A **parent** is a living thing that makes offspring. Adult plants can make more plants. These adult plants are called parents.

Offspring are the young made by parents. A young plant is also called a seedling. Seedlings grow into adult plants. Seedlings inherit some things from their parents.

🖱 GO ONLINE

INTERACTIVE PRESENTATION

Close Reading: Offspring are Like Their Parents

A C T   **Access Complex Text**

Connection of Ideas Tell students that when they read a complex text, they should try to connect the information they read to what they have already learned. Have students recall what they have already learned about plant development.

• **ASK:** What structures do young plants inherit from their parents? Sample answer: Young plants inherit the type of fruit they make from their parents.

• **ASK:** Can flower color be used to tell if two plants are related? Sample answer: No. Some plants will make different colored flowers than their parents.

Make Connections

Talk About It

What else do you think offspring inherit from their parents?

Notes

_____

_____

_____

Inherit is when something is passed from the parent to its young. A plant inherits its type of leaves or fruit. Offspring are like their parents. <u>Offspring and their parents have the same structures.</u> Offspring are also different than their parents. Their flowers may be a different color.

Look at the photo of the tulips.

✓ how the tulips are different.

 color

 type of flower

Explain **Lesson 1** Plants and Their Parents **61**

Make Connections

Talk About It

Encourage students to brainstorm a list of things young plants inherit from their parents.

Scientific Vocabulary

inherit Tell students the word inherit comes from the Latin word *heres* meaning "heir".

REVISIT

PAGE KEELEY SCIENCE PROBES

Have students return to the Page Keeley Science Probe. Students should revise their answer based on what they have learned throughout the lesson. Encourage discussion about how their opinions have changed about young plants and their parents.

FOLDABLES **Study Guide Foldables®**

🕙 10 min   👥 whole class

Have students make a Half-Book Foldables using page EM13-EM14 for guidance. Instruct students to write "Young Plants" on the front cover and "Parent Plants" on the back cover. Have students explain on the inside of the book how young plants and their parents are alike and different. Encourage students to draw pictures to help explain.

EL Support

ELD.PII.1.6, DCI.L1.A, DCI.LS3.A, DCI.LS3.B, SEP – 4, CCC-1 Guide students to connect ideas by summarizing how adult plants and their offspring are alike and different.

EMERGING

Support students in drawing an adult plant and its offspring and label words for how they are alike and different. Then provide them with a sentence frame to connect ideas using *and/but*: Adult plants have _____, and their offspring have _____.

EXPANDING

Support students in saying how adult plants and offspring are alike and different and show them on a Venn Diagram. Then help students use *because* to show cause/effect relationships: Adult plants have _____ [seeds] because they need to _____. [make offspring]

BRIDGING

Provide students with a Venn Diagram to list how adult plants and their offspring are alike and different. Then encourage them to combine ideas using *so*. EX. Adult plants are fully grown so they are bigger than their offspring.

INQUIRY ACTIVITY | Simulation

Plants Grow and Change

 Prep: 5 min I **Class:** 30 min   small groups

Purpose

Students will investigate how several species of plants change as they grow.

Before You Begin

Review the simulation prior to the activity and make notes on any points you would like to discuss.

Have students discuss their prior knowledge about how these plants change as they grow. Focus students on the question they will investigate:

ASK: How will the plants change as they grow?

Post the question to be able to revisit as a class during the activity.

Guide the Activity

Make a Prediction Help students predict changes they expect to see as plants grow and develop.

Investigate

Students may need help using the simulation.

3. Remind students to take photos of the plants as they grow. They will be used to compare later.

Short on Time?

If you are short on time, use this simulation as a teacher demonstration. This is a great opportunity to show students how to follow the steps of an investigation, collect data, and draw conclusions. Ask students to describe what they observe as you draw the pictures of the plant at various times. Demonstrate the skill of analyzing and interpreting data as you use compare your results with your prediction.

INInquiry box

INQUIRY ACTIVITY

Simulation

Plants Grow and Change

Observe how green beans, carrots, corn, and peas grow into adult plants.

 GO ONLINE

Use the simulation *Plants Grow and Change* to learn about how plants develop.

Make a Prediction How will the young plants change as they grow?

Sample answer: The young plants will grow larger.

_____

Investigate

1. Plant each vegetable in the garden.

2. Watch the plants grow.

3. Take photos as the plants grow.

4. **Record Data** Choose one plant and draw how it changes over time.

62   Explain   **Module:** Plant Parents and Their Offspring

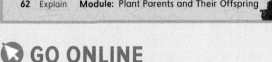 GO ONLINE

INTERACTIVE PRESENTATION

Inquiry Activity: Plants Grow and Change

INQUIRY ACTIVITY

Identifying Preconceptions

Point out that nearly all plants appear different from their parents while they are young. After plants are fully grown, they may still appear different from their parents. It is important that students observe how plants change as they grow to better determine if differences they observe between plants and their offspring are due to inheritance of traits or stage of development.

Plant	2 Weeks	4 Weeks	12 Weeks
Sample answer: carrot	Drawing could include a thin stem, two thin leaves, and stringy roots.	Drawing could include a thicker stem with more leaves. Below the ground a thin carrot is growing with stringy roots.	Drawing could include a thicker stem with more leaves. Below the ground is a thicker carrot with some stringy roots.

Communicate

5. Did your observations match your prediction?

Sample answer: Yes. The young plant grew larger.

Think about the patterns you observed. Circle the structures that seedlings do not have. Explain your answer to a partner.

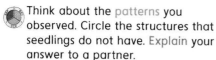

Explain   **Lesson 1** Plants and Their Parents   **63**

SEP Constructing Explanations and Designing Solutions

Students should use evidence to construct explanations about the phenomena they observe. As students progress, they will use this skill to design solutions to problems they encounter in their lives.

ASK: Explain how the plant you chose changed as it developed. Sample answer: The carrot stem grew larger, more leaves developed, and the carrot formed underground.

ASK: A farmer has a problem. He just found a young plant he does not know on his farm. How can he figure out what type of plant it is? Design a solution to help this farmer. Sample answer: The farmer could make observations as the plant grows. Then he could do research to find other plants that look similar to the one he found.

Communicate

Students should select one plant to use for data collection. Students can enlarge each photo by clicking on it or returning to the main screen.

5. Help students determine that young plants do not have corn, green beans, pea pods, or carrots. These structures do not develop until later in the plant life cycle.

Inquiry Spectrum

🔵 **GO ONLINE** for guidance on how to adapt this activity to a different level of inquiry.

Three-Dimensional Thinking

SEP: Constructing Explanations and Designing Solutions

CCC: Patterns

Check students' notebooks for accuracy. Students should be able to identify a structural pattern between seedlings of various plants. Have students explain their answers.

Formative Assessment

Use this opportunity to do a quick assessment to determine whether students are ready to move on. You may choose to assess students as a whole group or individually using the exit slip strategy.

ASK: What does a young plant inherit from its parents? Sample answer: The young pea plant inherited its color and leaf shape from its parent.

INQUIRY ACTIVITY | Hands On

Grow a Radish

 Prep: 10 min | **Class:** 30 min   small groups

Purpose

Students will observe how a radish changes as it grows and develops.

Materials

Additional: plastic gloves, cubes

Before You Begin

Discuss children's prior experience with growing plants and with radishes, specifically. Focus students on the question that they will investigate:

ASK: What will an adult radish plant look like?

Post the question to be able to revisit as a class during the activity.

Guide the Activity

Make a Prediction Encourage students to draw a picture of what they believe an adult radish will look like. Point out the photo of the sliced radish on the page to get students thinking.

Investigate

BE CAREFUL Make sure that students wear nonallergenic gloves when handling the radish plants.

1. Grow multiple radishes in the event that some seeds do not germinate, so students will still be able to make observations. Refer to the seed packet for instructions on proper growing conditions.

4. It will take approximately 4 weeks for the radishes to grow. Set aside 10 minutes each week for students to make and record observations. Be sure to harvest the plant at 4 weeks to observe the underground structures.

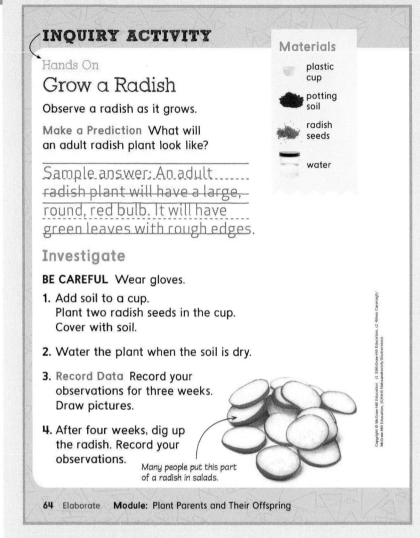

INQUIRY ACTIVITY

Hands On

Grow a Radish

Observe a radish as it grows.

Make a Prediction What will an adult radish plant look like?

Sample answer: An adult radish plant will have a large, round, red bulb. It will have green leaves with rough edges.

Materials
- plastic cup
- potting soil
- radish seeds
- water

Investigate

BE CAREFUL Wear gloves.

1. Add soil to a cup. Plant two radish seeds in the cup. Cover with soil.

2. Water the plant when the soil is dry.

3. **Record Data** Record your observations for three weeks. Draw pictures.

4. After four weeks, dig up the radish. Record your observations.

Many people put this part of a radish in salads.

64 Elaborate   **Module:** Plant Parents and Their Offspring

🖱 GO ONLINE

INTERACTIVE PRESENTATION

Inquiry Activity: Grow a Radish

INQUIRY ACTIVITY

Week 1	Week 2
Drawing should include one or two small leaves that appear just above the soil.	Drawing should include a short, thin stem and two or three leaves that appear larger.

Week 3	Week 4
Drawing should include multiple leaves and a stronger stem. The bulb may begin to show.	Drawing should include an adult radish plant with roots; a large, round, red bulb; and a stem with green leaves with rough edges.

Communicate

5. Did what you observed match your prediction? Explain.

Yes. The adult radish plant has a red round bulb. A stem and green leaves with rough edges.

Communicate

5. Have students compare their results with their predictions. Help students determine whether or not their predictions were supported by their observations.

You may wish to have students write about how the radish grew.

Short on Time?

Photos of a young radish can be found online or in a book. Adult radishes can be purchased at a local grocery store to allow for hands-on observations.

MATH ▶Connection

Measurement and Data 1.MD.A.2

Have students use measuring cubes to measure the height of their radish plant as it grows. Encourage students to record their data in the data table.

Inquiry Spectrum

Structured Inquiry

This is a structured inquiry activity, students are given a question to investigate and procedure to follow.

Guided Inquiry

Ask students to cover up the steps of the investigation. Present students with the objective. Ask them to develop investigation steps.

Open Inquiry

Show students a radish seed and an adult radish plant. Have students ask questions about the seed and the adult radish. Allow students to develop their own investigations to answer their questions.

LESSON 1 REVIEW

 20 min   whole class

EXPLAIN THE PHENOMENON

Have students revisit the photo of the dandelion as they answer the Explain the Phenomenon question. You may want to show the *Dandelion Seeds* video again.

| **Revisit the Phenomenon** question:

Are these plants related?

➡ This led to the overarching lesson **Essential Question:** How are plants similar to and different from their offspring?

Encourage students to review the notes and questions they wrote. Students should try to answer the questions that they had at the beginning of the lesson.

REVISIT

Students should revisit the Page Keeley Science Probe to decide whether they would like to change or justify their response. Students have had an opportunity to develop a conceptual understanding of plants and their parents. Revisiting the probe here will reveal whether students are holding onto a misconception or have gaps in conceptual understanding.

🔄 **GO ONLINE** to explore the Vocabulary Flashcards with students to review lesson vocabulary.

LESSON 1

Review

EXPLAIN THE PHENOMENON | Are these plants the same?

Summarize It

How are adult plants and their offspring alike and different?

Sample answer: Adult plants and their offspring both have leaves, stems, and roots. Adult plant have flowers which make seeds.

REVISIT PAGE KEELEY SCIENCE PROBES | Look at the Page Keeley Science Probe on page 51. Has your thinking about young plants changed?

66   Evaluate   **Module:** Plant Parents and Their Offspring

🔄 GO ONLINE

INTERACTIVE PRESENTATION	INTERACTIVE PRESENTATION
Lesson Review: Plants and Their Parents	Vocabulary Flashcards: Plants and Their Parents
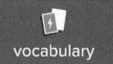	

Three-Dimensional Thinking

Answer the questions based on what you learned about plants.

1. Which plant will this seedling grow into?

(Circle) the adult plant.

Copyright © McGraw-Hill Education. (l to r) to Ingipad/Shutterstock, Zeljko Radojko/mock/Getty Images. Mariaktsa Wittov/Getty Images

2. Offspring are exactly like their parents.

☐ true

✓ false

Evaluate **Lesson 1** Plants and Their Parents **67**

🔽 GO ONLINE

ADDITIONAL RESOURCE

Lesson Check: Plants and Their Parents

ADDITIONAL RESOURCE

Vocabulary Check: Plants and Their Parents

vocabulary

LESSON 1 REVIEW

 ## Three-Dimensional Thinking

Have students apply their three-dimensional learning to show their understanding.

1-LS3-1: Make observations to construct an evidence-based account that young plants and animals are like, but not exactly like, their parents.

1. Students should have circled the photo of the corn. **LS3.A, CCC-1, DOK-2**

2. This statement is false. Offspring are similar to, but not exactly the same as their parents. **SEP-6, LS3.B, CCC-1, DOK-2**

Online Assessment Center

You might want to assign students the lesson check that is available in your online resources. You can assign the premade lesson check, which is based on the Disciplinary Core Ideas for the lesson, or you can customize your own lesson check using the customization tool.

🔽 **GO ONLINE** explore the Vocabulary Check: *Plant Parts* with students to review the vocabulary from the lesson or assign to students to evaluate their lesson vocabulary knowledge.

LESSON 1 REVIEW

Extend It

 15 min   partners

This task focuses gives students the opportunity to engage in an open inquiry while also building the 21st Century Skill of creativity. Have students plan an investigation to learn how their plant will change as it grows.

Extend It Scoring Rubric

Use the following rubric guidelines to assess the Extend It activity.

3 Points The student selects a plant, designs a plausible investigation, and includes a labeled diagram to help explain his or her design.

2 Points The student selects a plant, designs an investigation, and includes a labeled diagram to help explain his or her design. However, either the investigation or diagram has errors or inaccurate information.

1 Point The student selects a plant, designs an unworkable investigation, and includes a labeled diagram with inaccuracies.

Extend It

A friend wants to know how plants change as they grow. Choose a plant. Design an investigation to show how your plant will change as it grows. Write or draw your design below. Use labels.

Drawings could include a student observing a young plant over several weeks.

68   Evaluate   **Module:** Plant Parents and Their Offspring

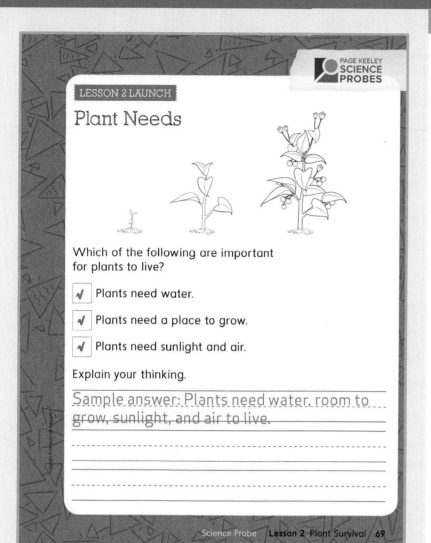

GO ONLINE

INTERACTIVE PRESENTATION

Science Probe: Plant Needs

SCIENCE PROBES

PAGE KEELEY
SCIENCE PROBES

Plant Needs

🕙 10 min   👥 whole class

Using the Probe

The purpose of this probe is to identify students' prior knowledge about plant needs. This probe works well with a confidence level assessment. Have students think about their answers and mark them in their notebook. Read each answer choice, and ask students to show how confident they are in their answer by raising 1, 2, or 3 fingers. Use this probe to assess students' prior knowledge of the lesson content and to identify possible misconceptions.

Be sure not to tell students the answer. It is not important that students know the answer to this probe at this point in the lesson. What is important is the reasoning they provide to support their answer. Students will revisit the probe throughout the lesson to see how their thinking has changed.

Throughout the Lesson

Use students' explanations to bridge the students' initial ideas about plant needs with the understanding they develop. Prompts in the Teacher's Edition will instruct you when it's time for students to revisit the probes.

Teacher Explanation

Plants need water, a safe space to grow, sunlight, and air (oxygen) to live. Students should have marked all the answer choices.

Teacher Toolbox

Identifying Preconceptions

This probe is designed to reveal if students can recognize the basic needs of plants. Students may not realize that all plants need a sunlight and air to live. They may think that plants underwater don't use air. These students do not realize that underwater plants use the oxygen in the water for important life functions. Likewise, plants need sunlight to perform photosynthesis.

Building to the Performance Expectations

In this lesson, students will explore content and develop skills leading to mastery of the following Performance Expectations:

K-2-ETS1-3. **Analyze data from tests of two objects designed to solve the same problem** to compare the strengths and weaknesses of how each performs.

1-LS1-1. **Use materials to design a solution to a human problem by mimicking how plants and/or animals** use their external parts to help them survive, grow, and meet their needs.

SEP Science and Engineering Practices

Analyzing and Interpreting Data

Analyzing data in K-2 builds on prior experiences and progresses to collecting, recording, and sharing observations. (K-2-ETS1-3)

Constructing Explanations and Designing Solutions

Constructing explanations and designing solutions in K–2 builds on prior experiences and progresses to the use of evidence and ideas in constructing evidence-based accounts of natural phenomena and designing solutions. (1-LS1-1)

DCI Disciplinary Core Idea

ETS1.C: Optimizing the Design Solution

Because there is more than one possible solution to a problem, it is useful to compare and test designs. (K-2-ETS1-3)

LS1.A: Structure and Function

All organisms have external parts. Different animals use their body parts in different ways to see, hear, grasp objects, protect themselves, move from place to place, and seek, find, and take in food, water and air. Plants also have different parts (roots, stems, leaves, flowers, fruits) that help them survive and grow. (1-LS1-1)

LS1.D: Information Processing

Animals have body parts that capture and convey different kinds of information needed for growth and survival. Animals respond to these inputs with behaviors that help them survive. Plants also respond to some external inputs. (1-LS1-1)

CCC Crosscutting Concept

Influence of Science, Engineering and Technology on Society and the Natural World

Every human-made product is designed by applying some knowledge of the natural world and is built using materials derived from the natural world. (1-LS1-1)

Structure and Function

The shape and stability of structures of natural and designed objects are related to their function(s). (1-LS1-1)

ELA/Literacy Connections

RI.1.3 Reading, W.1.8 Writing

Math Connections

1.MD.A.2 Measurement and Data

* See correlation table for full text of ELA and Math standards.

Track Your Progress to the Performance Expectations

You may want to return after completing the lesson to note concepts that will need additional review before your students start the module Performance Project.

Dimension	Concepts to Review Before Assessment
SEP Analyzing and interpreting Data (K-2-ETS1-3)	
SEP Constructing Explanations and Designing Solutions (1-LS1-1)	
DCI ETS1.C: Analyzing and Interpreting Data (K-2-ETS1-3)	
DCI LS1.A: Structure and Function (1-LS1-1)	
DCI LS1.D: Information Processing (1-LS1-1)	
CCC Influence of Science, Engineering and Technology on Society and the Natural World (1-LS1-1)	
CCC Structure and Function (1-LS1-1)	

Lesson at a Glance

Full Track is the recommended path for the complete lesson experience. FlexTrack A
and FlexTrack B provide timesaving strategies and alternatives.

	Day-to-Day	Pacing	**Full Track** 45 min/day (full year) Resources
Assess Prior Knowledge	Page Keeley Science Probe: *Plant Needs*	Day 1	Page 69
Engage	Discover the Phenomenon: What happened to this plant?		Pages 70—71 Video: *Changing Flower*
Explore	Inquiry Activity: *Plants and Shade*	Day 2	Pages 72—73
	Science Read Aloud: *A Little Seed's Journey*		Page 74
Explain	Make Your Claim: *Do plants need the same things to stay alive?*	Day 3	Page 75
	People Can Help Plants Science Read Aloud: *Making New Plants*		Pages 76—77 Video: *How Plants Survive*
	Close Reading: *Seeds Move*	Day 4	Pages 78—79 Digital Activity: *Seeds Move From Place to Place*
Elaborate	Survival Structures	Day 5	Page 80
	STEM Career Connection: *What Does a Horticulturist Do?*		Page 81
	Inquiry Activity: *Plant Survival*	Day 6	Pages 82—83
Evaluate	Explain the Phenomenon: What happened to this plant?	Day 7	Pages 84—89
			7 Days

Essential Question: How do plants survive?

Objective: Explain how plants use their structures to meet their needs and survive.

Vocabulary: need, pollen, survive

FlexTrack A	
30 min/day (5 days per week)	
Pacing	**Resources**
Day 1	Page 69 Employ the Sticky Bar Graphs strategy.
	Pages 70—71 Video: *Changing Flower*
Day 2	Pages 72–73 Use the simulation as a teacher demonstration
Day 3	
	Pages 76–77 Video: *How Plants Survive*
Day 4	Pages 78–79
Day 5	Page 80
Day 6	Pages 82–83 Have appropriate research materials on hand for students to read. Provide easy-to-read resources, such as one-page summary sheets printed from trusted Internet sites or a few bookmarked pages in a nonfiction book, that students can use to research their chosen plant.
Day 7	Pages 84–86 Answer the Explain the Phenomenon question as a class.
7 Days	

FlexTrack B	
30 min/day (3 days per week)	
Pacing	**Resources**
Day 1	Page 69 Employ the Sticky Bar Graphs strategy.
	Pages 70—71 Video: *Changing Flower*
Day 2	Pages 72–73 Use the simulation as a teacher demonstration
Day 3	Video: *How Plants Survive*
	Pages 78–79
Day 4	Pages 84–86 Answer the Explain the Phenomenon question as a class.
4 Days	

Lesson 2: **Plant Survival**

ENGAGE   EXPLORE   EXPLAIN   ELABORATE   EVALUATE

Lesson Objective

Students will explore how plants use their parts to survive, grow, and meet their needs. Students will observe plant structures and functions to construct explanations about how plants survive.

DCI **Optimizing the Design Solution**

ETS1.C Becuase there is always more than one possible solution to a problem, it is useful to compare and test designs.

DCI **Structure and Function**

LS1.A All organisms have external parts. Different animals use their body parts in different ways to see, hear, grasp objects, protect themselves, move from place to place, and seek, find, and take in food, water and air. Plants also have different parts (roots, stems, leaves, flowers, fruits) that help them survive and grow.

DCI **Information Processing**

LS1.D Animals have body parts that capture and convey different kinds of information needed for growth and survival. Animals respond to these inputs with behaviors that help them survive. Plants also respond to some external inputs.

LESSON 2

Plant Survival

70   Module: Plant Parents and Their Offspring

Teacher Toolbox

Science Background

Plants have basic structures that function to keep the plant alive and healthy. In this lesson, students will learn about how plants function to survive as a species and as individuals. Throughout the lesson, help students make connections to how plant structures and functions help them plants overcome problems that could impact their survival.

DISCOVER
THE PHENOMENON

What happened to this plant?

GO ONLINE

Watch the video *Changing Flower* to see the phenomenon in action.

Look at the photo. Watch the video.
How do plants stay alive?
What did you observe?

Sample answer: This plant looks wilted. It might not have gotten enough water. Plants need water to live and grow.

Did You Know?
The longest living plant in North America is a bristlecone pine. It is more than 5,000 years old!

Engage   **Lesson 2** Plant Survival   **71**

GO ONLINE

INTERACTIVE PRESENTATION

Discover the Phenomenon:
Plant Survival

DISCOVER THE PHENOMENON

 10 min   whole class

Recall that scientists refer to an event or situation that is observed or can be studied as a phenomenon. Have students study the photo of the wilted plant.

Ask the **Discover the Phenomenon** question:

What happened to this plant?

This leads to the overarching module **Essential Question**:
How do plants use their parts to meet their needs?

GO ONLINE Check out *Changing Flower* to see the phenomenon in action.

Talk About It

Ask students to describe what they see. Help students turn their observations from the video into questions. Start a class discussion with the following prompts:

ASK: Why do plants need to survive?

ASK: Do people have any of the same needs as plants? Explain.

Record responses and questions on the board or chart paper to refer to as you move through the module.

Did You Know?

The oldest Great Basin bristlecone pine, named *Pinus longaeva*, is the oldest non-clonal organism on Earth. It is estimated to be 5,060 years old.

Lesson 2: **Plant Survival**

INQUIRY ACTIVITY | Simulation

Plants and Shade

 Prep: 5 min | **Class:** 30 min   small groups

Purpose

Remind students of the photo of the wilted plant they saw at the beginning of this lesson. Tell students they will investigate whether or not all plants need the same amount of sunlight.

Before You Begin

Reserve the technology cart so that students can use the online simulation. Have students think about how plants grow in different amounts of sunlight. Focus students on the question that they will investigate:

ASK: How will the amount of sunlight affect the growth of a plant?

Post the question so that you can to be able to revisit it as a class during the activity.

Guide the Activity

Make a Prediction Help students make a prediction about sunlight and plant growth based on their previous knowledge. Remind students that they have encountered many types of plants growing in various locations. They may have seen grass growing next to a sidewalk, flowers in a park, or trees along the side of a road. Have students think about the amount of sunlight each of these plants receives.

Investigate

1-3. Help students identify the blue and white flowers.

4. Make sure students can identify the part of the yard that gets the most sunlight. Ask students to compare the blue and white flowers to determine which grew best in sunlight. Have students repeat the simulation if necessary.

Inquiry Spectrum

 GO ONLINE for guidance on how to adapt this activity to a different level of inquiry.

INQUIRY ACTIVITY

Simulation

Plants and Shade

Plants cannot grow if they do not get what they need. Observe how two different plants grow in shade and sunlight.

GO ONLINE

Use *Plants and Shade* to see how plants grow.

Make a Prediction How will the amount of sunlight affect the growth of a plant?

Sample answer: Different plants need different amounts of light. If a plant does not get enough sunlight, it will turn brown. The plant will not survive.

Investigate

1. Plant three blue flowers in each part of the yard.

2. Observe what happens each year.

3. Repeat the simulation with the white flowers.

72 Explore   **Module:** Plant Parents and Their Offspring

🡒 GO ONLINE

INTERACTIVE PRESENTATION

Inquiry Activity: Plants and Shade

INQUIRY ACTIVITY

Communicate

4. Draw a picture of the plant that grew best in the sunlight.

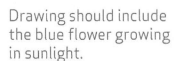

Drawing should include the blue flower growing in sunlight.

Communicate

5. Do all plants need the same amount of light? Use your observations to explain.

Sample answer: No. The blue flower grew best in full sunlight. The white flower grew the best in the shade.

 Talk About It

Do your observations match your prediction? Tell a partner.

SEP Constructing Explanations and Designing Solutions

Students in first grade should be collecting evidence to construction explanations. This will progress to students using their explanations to design solutions to problems they encounter.

ASK: A friend is trying to grow a plant. She puts the plant in the shade, but it is not growing well. How could you use what you learned in this simulation to help your friend? Sample answer: I learned that plants need different amounts of sunlight to grow. I would help my friend by telling her to grow her plant in a place with more sunlight.

ASK: What do you think farmers need to know about plants and sunlight? Sample answer: Farmers need to know how much sunlight each plant needs to grow.

Communicate

Have students write a conclusion about the amount of light needed by plants. If time allows, have students observe what happens to the pink flower and explain whether or not their observations about this flower support their conclusion.

Short on Time?

If you are short on time, use this simulation as a teacher demonstration. Use this as an opportunity to show students how to progress through an investigation. Ask students to explain which plant grew best in the sunlight. Demonstrate the skill of constructing an explanation about the relationship between plants and sunlight.

Talk About It

Have partners revisit their predictions from the beginning of this activity and determine whether or not they were supported by their observations from the simulation.

Differentiated Instruction

AL In response to the Inquiry Activity, ask students to draw a picture of a plant in a healthy environment with all the necessities: sun, water, and nutrients.

OL Have partners plan a similar experiment that they could do at home. Have them use a word wall to help them identify what they will need to conduct the experiment. Then have them make predictions about outcomes.

BL Have students construct an explanation about why some plants in the activity were unhealthy. Have them write at least two sentences.

Seeds

 20 min   partners

Before Reading

Have students observe the picture on page 4 of the Science Paired Read Aloud *A Little Seed's Journey*. Ask students to share what they know about seeds. Have students make a list of seeds they know on the board.

Science Paired Read Aloud

Read aloud with students pages 4-13 from the Science Read Aloud *A Little Seed's Journey*. Students will learn how seeds get from one place to another.

After Reading

Ask students to summarize what happened in the reading. Ask students to relate what they read to the module phenomenon of the dandelion blowing in the wind.

ASK: How did the structure of the fruit help the seed travel to a new place? Sample answer: The spiky skin of the fruit stuck to the raccoon.

ASK: What do you think one of the functions of the spikes on the spiky fruit is? Sample answer: I think one of the functions of the spikes is to stick to things.

CCC Structure and Function

Remind students what they have learned about structure and function. Encourage students to think about how a thing's structure helps it function. Obtain a photo of a microscopic view of a hook-and-loop fastener. Ask students to think about how the structure of the fasteners help them function.

 **GO ONLINE** Have students listen to the Crosscutting Concept Science Song: *Structure and Function*.

Crosscutting Concepts Graphic Organizer

 **GO ONLINE** Use the Crosscutting Concept Graphic Organizer to determine the relationship between a plant structure and its function.

Seeds

All plants need sunlight to grow. Some plants need more sunlight. Some plants need less sunlight. Seeds need things too. Seeds need water and a place to grow.

📖 Listen to *A Little Seed's Journey*.

1. How did the little seed travel to a new place?

The seed stuck to the fur of a raccoon.

2. What do you think will happen to the little seed?

Sample answer: The seed will grow into a new plant.

 GO ONLINE

INTERACTIVE PRESENTATION

Read Aloud: A Little Seed's Journey

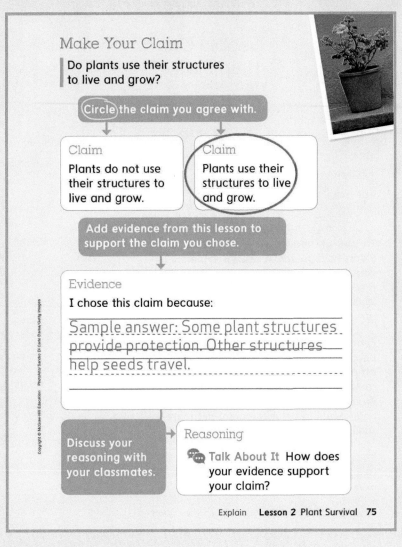

Make Your Claim

Do plants use their structures to live and grow?

(Circle) the claim you agree with.

Claim
Plants do not use their structures to live and grow.

Claim
Plants use their structures to live and grow.

Add evidence from this lesson to support the claim you chose.

Evidence

I chose this claim because:

Sample answer: Some plant structures provide protection. Other structures help seeds travel.

Discuss your reasoning with your classmates.

Reasoning

Talk About It How does your evidence support your claim?

Explain **Lesson 2** Plant Survival **75**

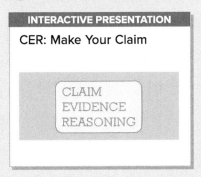

GO ONLINE

INTERACTIVE PRESENTATION

CER: Make Your Claim

CLAIM
EVIDENCE
REASONING

MAKE YOUR CLAIM

 15 min   individual

Students will be making a claim that answers a question about plant needs. Students will support their claim with evidence from the Simulation: *Plants and Shade*. As the lesson continues, students will revisit this page to add evidence. Students will use scientific knowledge gained throughout the lesson to provide reasoning that supports their claim.

CLAIM

Give students time to reflect and brainstorm possible answers. Have students circle their answers in the graphic organizer. A student's claim should take a clear stand on whether all plants have the same needs.

EVIDENCE

Scientific evidence is information that supports or contradicts a claim. This information can come from a variety of sources. Research, experimentation, or data interpretation are common sources of scientific evidence. Students can provide evidence based on what they observed in the Inquiry Activity. If students find that their evidence does not support their claim help them determine why this happened. The students might have collected incorrect data, misinterpreted their results, or selected the incorrect claim initially. It is also possible that students have not yet gathered enough evidence and must continue exploring the content. Throughout this lesson, encourage students to return to their claim to add more evidence. Look for the blue square for a reminder to revisit the Claim, Evidence, and Reasoning page.

REASONING

When providing reasoning, students must explain the scientific knowledge, principle, or theory they used to support their argument. If, for example, a student claims that plants have different needs, they must provide a scientific explanation that explains why plants need different things.

People Can Help Plants

 40 min   pairs

VOCABULARY

Encourage students to use context clues to derive the meaning of vocabulary words. Struggling students can use the glossary to access the definitions.

GO ONLINE to watch the video *How Plants Survive* individually or play the video for the whole class.

Scientific Vocabulary

survive Explain to students that the word survive has Latin origins. The Latin word vivere means "live".

need Remind students that the word need is often used as a verb, meaning "to require something." They may have even said themselves, "I need this." Discuss the similarities and differences of the multiple meanings of the word need.

PRIMARY SOURCE

This photo was taken in 1942 in Imperial County, California. It shows men working on a farm to care for tomato plants. This photo comes from the Library of Congress.

Differentiated Instruction

AL Draw a Venn diagram on the board, and help students compare and contrast what plants and humans need for survival.

OL Have partners list the ways that plants protect themselves. Then have them discuss how humans mimic plants in the way that they protect themselves.

BL Have partners research how a plant, such as the Venus flytrap or the rose, has adapted over time. For example, have them construct an explanation for why a rose has thorns.

Teacher Toolbox

Science Background

Plants must meet their needs to survive. Some needs are universal such as sunlight, water, and oxygen. All plants need water and oxygen to grow. Other needs vary. Plants have different sunlight requirements, use various ways to collect nutrients, and employ different methods to move their seeds and spores to new locations.

Vocabulary

Look and listen for these words as you learn about how plants stay alive.

need

pollen

survive

People Can Help Plants

PRIMARY SOURCE

People care for tomato plants in Imperial County, California in 1942.

To **survive** means to live and grow. Plants have needs. A **need** is something you must have in order to live. Water is a need because plants must have water to live and grow. Plants also need air and space to grow.

People can help plants grow. Sometimes people take leaves, stems, or branches off plants. This gives the plants more room to grow. This is called pruning. Pruning helps plants get what they need. Pruning helps the plant get more sunlight. It can make fruits grow larger.

GO ONLINE

Watch the video *Seeds* to learn more about how plants grow.

76 Explain   **Module:** Plant Parents and Their Offspring

GO ONLINE

INTERACTIVE PRESENTATION	ADDITIONAL RESOURCE
Read About: People Can Help Plants	Lesson Vocabulary: Plant Survival
	 vocabulary

Listen to *Making New Plants.*

Pollen is a powder that helps make new plants. Plants that have flowers or cones need pollen to make offspring. Pollen is very light. It can be spread by the wind or by animals. Some plants attract animals using color and smell.

1. How do animals and people help plants survive?

   Sample answer: People prune plants to help them grow. Animals carry seeds so new plants can grow.

2. How do plants get animals to spread their pollen?

   Sample answer: Plants attract animals using their color and smell. Animals spread pollen from plants on their bodies.

Explain   **Lesson 2** Plant Survival   77

🖱 GO ONLINE

INTERACTIVE PRESENTATION

Read Aloud: Comparing Plant Parts

Before Reading

Have students observe the picture on page 14 of the Science Paired Read Aloud *Making New Plants*. Ask students to share what they know about the structure and function of seeds.

📖 Science Read Aloud

Read aloud with students pages 14-23 from the Science Paired Read Aloud. Students will learn how seeds create new plants. Students will encounter the vocabulary word: *pollen*. Discuss the questions on page 24 as a class.

After Reading

Have students compare the fiction and nonfiction texts. Ask how these two texts are similar. Have students discuss how seeds help plants find new places to grow.

READING Connection

Reading Informational Text RI.1.3

As students answer the questions about the nonfiction text, encourage them to make connections to the fiction reading and their own personal experiences. Help students identify the key ideas and details from the text.

EL Support

ELD.PI.1.1, SEP-1 Before showing the *How Plants Survive* video, review the following terms: *seed, water, sunlight, plant, color, scent, insect, pollinate, cycle*. Have students exchange ideas about which of these things a plant needs to grow: A plant needs sunlight. A plant does not need color.

EMERGING
Show the video once, and then start it again. This time, pause the video at certain points and have students use vocabulary to name what they see.

EXPANDING
Have partners discuss the video and tell what they can remember. Then have them watch again and add to their list: I saw an insect. I saw a seed.

BRIDGING
Have partners think of questions about the video. Then have them turn to another student and ask and answer questions.

(CLOSE) READING

Seeds Move

 30 min   partner

Inspect

Read Have students read the passage to focus on understanding the overall meaning. Ask students to take notes to gather information about the ways seeds move.

Find Evidence

Have students reread the text. Remind students to underline the evidence that helps answer the question:

ASK: Do all plants have the same type of seeds? Explain.
Sample answer: Plants have different types of seeds. I learned that some seeds float on water and others can float on air.

A C T   Access Complex Text

Prior Knowledge Tell students that they may have already had experiences observing seeds moving. Students may have seen dandelion or maple seeds traveling through the air. Or students may have come across animals or humans eating the seeds from nuts, flowers, or fruits.

• **ASK:** Where have you seen plant seeds traveling before?
Sample answer: I have seen a coconut floating on the water on my favorite televison show.

• **ASK:** Where have you seen people or animals eating seeds?
Sample answer: I have seen squirrels eating seeds outside. I have seen people eating sunflower seeds at a sports game.

Differentiated Instruction

AL Have students research and then draw a landscape scene to illustrate the different ways that seeds are dispersed.

OL Have students explain what would happen with the model if there was no wind or little wind to carry the seeds away.

BL Have students choose a plant and make up a story with the plant as the main character. The story should include information about how the plant reproduces. Challenge them to use words from a word wall in their stories.

(CLOSE) READING

Inspect

Read the passage *Seeds Move.* Circle ways that seeds can travel.

Find Evidence

Do all plants have the same type of seeds? Underline the text that helps answer this question.

Notes

_____

_____

_____

_____

Seeds Move

Plants cannot grow in the same place as their parents. Plants have needs. Young plants need their own air, water, and soil. Seeds do not have legs, so they get around in other ways. Some seeds travel in water. These seeds can usually float. Some seeds travel by wind. These seeds are usually small or have parts that work like wings.

Make Connections

🗨 Talk About It

Tell a partner about a traveling seed you have seen.

Some seeds are eaten by (animals) and dropped in new places. Other seeds get caught on animals. These seeds usually have a way to stick to things.

Draw a picture of a traveling seed.

> Drawing could include a dandelion seed blowing in the wind.

Make Connections

Talk About It

Have partners or small groups think about plant seeds they have seen. Students should use descriptive words to explain the seeds they have seen.

 **REVISIT** Have students return to the Science Probe. Students should revise their answer based on what they have learned during this lesson. Encourage discussion about how students' opinions differ about plant needs.

■ COLLECT EVIDENCE Have students revisit their claims and add evidence and reasoning to the chart from the beginning of this lesson.

Examples of evidence include: quotations from a reading that states plants can have different structures that help them meet their different needs.

▶ **GO ONLINE** and use *Seeds Move from Place to Place* to learn how seeds travel.

Formative Assessment

Use this opportunity to do a quick assessment to determine whether students are ready to move on. You may choose to assess students as a whole group or individually using the exit slip strategy.

ASK: How does the structure or function of a seed help it move? Sample answer: A seed that travels by getting eaten by an animal will taste good to animals.

⏱ Time to Move

Instruct students to act out the ways that a seed can travel. Turn this into a game by having one student perform at a time. Encourage students to raise their hands and guess what type of seed is being shown.

Survival Structures

 30 min   whole class

Have students observe each photo and read the caption. Many plants have structures that help them survive. Plants like the dormilona respond to external stimuli. This plant has the ability to process information.

 ENGINEERING Connection

Engineering Design K-2-ETS1-1

Have students brainstorm ways that plants use their structures to meet their needs. Make a list on the board. Help students identify a human problem they can solve inspired by the structures of a plant. They may choose to use the student notebook for inspiration. Perhaps students want to find a way to stick two things together, move water, or keep animals out of a garden. Tell students to use what they have learned about plant survival to solve their problem.

💬 Talk About It

Have students find a partner. Tell students to explain what problem they solved. Then have them explain how their solution related to plants.

Leveled Reader

Use the Leveled Reader *How Plants Survive.* This book describes the parts of plants. Have students read the book with a partner. After reading, students can describe how parts of plants help them survive.

ASK: What plant parts help keep plants safe? Sample answer: Some plants have thorns to keep animals away. Other plants use smell to stay safe from animals.

CCC Influence of Science, Engineering and Technology on Society and the Natural World

Students should recognize that all man-made products are made from materials from the natural world. Similarly, designs are often heavily influenced by the natural world. Help students draw connections between the problems they identified and the plant-inspired solutions they developed.

Survival Structures

Plants have structures that help them survive, grow, and meet their needs.

Some plants have thorns or prickles. These sharp structures stop animals from eating the plant.

When the leaves of the dormilones plant are touched or shaken they fold inward.

The corpse flower releases a strong smell like rotting meat. This smell attracts insects. The insects spread its pollen.

Chestnut seeds grow inside prickly burrs. This structure protects the growing seed.

ENGINEERING Connection Copy the structures of a plant to solve a human problem. Explain the problem and draw a sketch. Use separate paper.

80 Explain   **Module:** Plant Parents and Their Offspring

🔵 GO ONLINE

INTERACTIVE PRESENTATION

Read About: Survival Structures

STEM CAREER Connection

What Does a Horticulturist Do?

STEM CAREER Connection
What Does a Horticulturist Do?

Horticulturists study plants. Horticulturists make sure plants and environments are healthy.

They help farmers grow better fruits and vegetables. Horticulturists work with farmers to protect the environment when they grow crops.

ENVIRONMENTAL Connection
Why is it important to humans that Earth has healthy plants? Explain on a separate sheet of paper.

> It is important that we take care of the environment.

Elaborate   **Lesson 2** Plant Survival   **81**

GO ONLINE

INTERACTIVE PRESENTATION

STEM Career: Horticulturist

STEM CAREER

STEM CAREER Connection
What Does a Horticulturist Do?

 10 min   whole class

Introduce the horticulturist STEM CAREER Connection. Encourage students to look at the photos and share their observations. Then ask students to summarize what a horticulturist does in their own words.

Begin a class discussion about how people can benefit from and alter natural systems. Make a list of natural systems (ex: universe, Earth, ocean) on a board or chart paper.

ASK: What might happen if there were no plants on Earth? Sample answer: Animals might not have the food they need.

ASK: How can people help keep natural systems like forests, parks, and gardens healthy? Sample answer: People can make sure they do not litter, take plants out of the soil, or bring new plants into these places without permission.

Academic Vocabulary

environment Tell students that the word *environment* comes from the French word *environ* which means to surround. Ask students to use context clues to determine what the word *environment* means. Develop a class definition then look up the word in a dictionary to compare.

Lesson 2 **Plant Survival** **81**

Lesson 2: **Plant Survival**

INQUIRY ACTIVITY | Research

Plant Survival

 Prep: 10 min | **Class:** 35 min   small groups

Purpose
Students will research mechanisms plants have that help them survive.

Before You Begin
Gather resources for students to use to conduct their research. Reserve a technology cart if needed. Show students pictures of plants growing in a desert or a tundra. Review with students what all plants need. Have students ask questions about the plants in the pictures or other plants they are interested in. Remind students throughout their research to focus their efforts on finding the answer to their question.

Guide the Activity

Ask a Question Help students choose a plant to investigate. Instruct students to write a question they would like to answer. This question is important because it will frame and guide the research.

Investigate

4. Have students record their answers to the questions they wrote on separate paper.

WRITING Connection

Research to Build and Present Knowledge W.1.8
Students should recall information they have gained throughout this lesson and conduct additional research to answer the question asked in this activity.

EL Support

ELD.LS1.A Use a Venn Diagram to review the things that plants need to survive and reinforce the idea that plants need different amounts of things. Have students use connecting words to compare two plants such as a cactus and radish.

EMERGING
Provide pictures of seeds, seedlings, and adult plants. Have students use *same/different* and *need/do not need* to tell about the two plants. Challenge them to use simple sentences: The seeds are different. Both plants need water.

EXPANDING
Have students compare and contrast using simple sentences. Both plants need sunlight. A cactus has spikes. A radish has lots of leaves.

BRIDGING
Encourage students to combine information into longer sentences. A radish needs lots of water, but a cactus needs very little water.

INQUIRY ACTIVITY

Research
Plant Survival
Some plants have special ways to stay alive. Research how plants survive.

Ask a Question Choose a plant. Write a question you will research.

Sample answer: How does a cactus survive in a desert?

Investigate

1. Use books and online resources to learn about your plant.

2. Find the answer to your question.

3. Draw a picture of the plant you chose. Label the structures that help the plant survive.

4. **WRITING** Connection Explain how your plant survives on a separate sheet of paper.

GO ONLINE

INTERACTIVE PRESENTATION

Inquiry Activity: Plant Survival

INQUIRY ACTIVITY

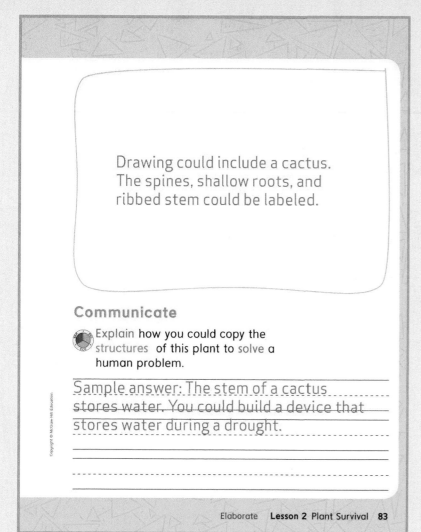

Drawing could include a cactus. The spines, shallow roots, and ribbed stem could be labeled.

Communicate

Explain how you could copy the structures of this plant to solve a human problem.

Sample answer: The stem of a cactus stores water. You could build a device that stores water during a drought.

Communicate

Students should present their results to the class. Each student should show the picture they drew of their plant and explain how their plant stays safe from danger.

Short on Time?

Direct students to investigate a specific plant. Provide students a question about the plant to further shorten the length of this activity. Gather materials students can use for their research.

Three-Dimensional Thinking

SEP: Constructing Explanations and Designing Solutions

DCI: LS1.A Structure and Function

Check student notebooks for accuracy. Ensure students have identified a problem that can be solved using inspiration from plant structures. Make sure student explanations offer plausible solutions.

Inquiry Spectrum

Structured Inquiry

Pose a question to students about plant survival, such as, How do lily pads survive in so much water? Direct students' research, giving them specific questions to answer.

Guided Inquiry

Give students a question to research for this activity, such as: How do cacti survive in the desert? Have students direct their own research.

Open Inquiry

This activity is open inquiry. You present a phenomenon to the class and students ask questions and investigate to find answers.

LESSON 2 REVIEW

 20 min   whole class

EXPLAIN THE PHENOMENON

Have students revisit the photo of the wilted plant as they answer the Explain the Phenomenon question. You may want to show the *Changing Flower* video again.

| **Rediscover the Phenomenon** question:
| What happened to this plant?

    This led to the overarching lesson **Essential Question**:
How do plants use their parts to meet their needs?

Encourage students to review the notes and questions they wrote. Students should try to answer the questions that they had at the beginning of the lesson.

 Students should revisit the Science Probe to decide whether they would like to change or justify their response. Students have had an opportunity to develop a conceptual understanding of plant survival. Revisiting the probe here will reveal whether students are holding on to a misconception or have gaps in conceptual understanding.

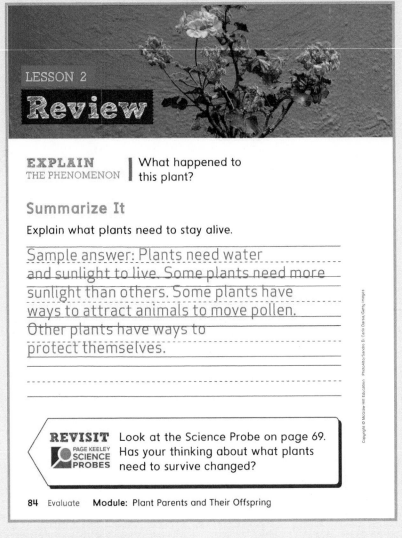

LESSON 2
Review

EXPLAIN THE PHENOMENON | What happened to this plant?

Summarize It

Explain what plants need to stay alive.

Sample answer: Plants need water and sunlight to live. Some plants need more sunlight than others. Some plants have ways to attract animals to move pollen. Other plants have ways to protect themselves.

REVISIT PAGE KEELEY SCIENCE PROBES | Look at the Science Probe on page 69. Has your thinking about what plants need to survive changed?

 **GO ONLINE** to explore the Vocabulary Flashcards with students to review lesson vocabulary.

GO ONLINE

INTERACTIVE PRESENTATION	**ADDITIONAL RESOURCE**
Lesson Review: Plant Survival	Vocabulary Flashcards: Plant Survival

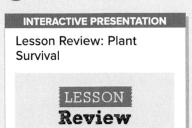

Three-Dimensional Thinking

Use what you have learned to answer the questions about plant survival.

1. What would happen if plants did not make seeds? Place a ✓ in the box if the statement is true.

☑ If plants did not make seeds, they could not make offspring.

☑ Some animals might also go hungry without seeds.

2. How does the shape of a flower help the plant survive?
Draw a model to help show what you know about structure and function.

Drawing should include a picture of a brightly colored flower, showing seeds or pollen.

Evaluate **Lesson 2** Plant Survival **85**

🐦 GO ONLINE

ADDITIONAL RESOURCE	ADDITIONAL RESOURCE
Lesson Check: Plant Survival	Vocabulary Check: Plant Survival

LESSON 2 REVIEW

 Three-Dimensional Thinking

Have students apply their three dimensional learning to show their understanding.

1-LS1-1 Use materials to design a solution to a human problem by mimicking how plants and/or animals use their external parts to help them survive, grow, and meet their needs.

1. Both answer choices are correct Students should have checked both answer boxes. **SEP-6, LS1.A, DOK-1**

2. Answers will vary. Sample answer: Drawings may include a colorful flower that attracts pollinators. Flowers may also contain seeds. **SEP-6, CCC-6, DOK-2**

Online Assessment Center

You might want to assign students the lesson check that is available in your online resources. You can assign the premade lesson check, which is based on the Disciplinary Core Ideas for the lesson, or you can customize your own lesson check using the customization tool.

🐦 **GO ONLINE** explore the Vocabulary Check: Plant Survival with students to review the vocabulary from the lesson or assign to students to evaluate their lesson vocabulary knowledge.

ENGAGE   EXPLORE   EXPLAIN   ELABORATE   **EVALUATE**

LESSON 2 REVIEW

Extend It

 15 min   partners

This task focuses on the 21st Century Skills of creativity and communication. Have students write a newspaper headline that describes the plant they investigated in the *Plant Survival* Inquiry Activity. Tell students to draw a picture. Remind students that a caption is a small piece of text that explains a what a picture or photo is showing. Have students write a caption to go with their picture.

Extend It Scoring Rubric

Use the following rubric guidelines to assess the Extend It activity.

3 Points Students include a headline, drawing, and caption that all relate to their research.

2 Points Students include only two of their following: a headline, drawing, and caption that all relate to their research.

1 Point Students include only one of the following: a headline, drawing, and caption that all relate to their research.

Extend It

Use your plant research to write a newspaper headline. Draw a picture to show what you learned. Write a caption to explain your picture.

Sample answer: Cacti Store Water

> Drawing could include a cactus showing its spines, shallow roots, and a ribbed stem.

Sample answer: The shallow roots quickly absorb water. The stem of a cactus stores water.

86   Evaluate   **Module:** Plant Parents and Their Offspring

STEM Module Project
Engineering Challenge

Design a Seed That Travels

A farmer wants to design a way to spread seeds across a farm. Use what you have learned. Design a model seed to help the farmer.

Build Your Model

1. List three ways seeds travel.

   (air)

   water

   animals

2. (Circle) how your seed will travel.

3. Draw your seed on the next page.

4. List the materials you will need.

5. Build and test your design.

6. **MATH > Connection** Measure how far your seed travels.

Materials
Sample materials:
construction paper
glue
tissue paper
tape

Look back at the lesson to see how seeds move!

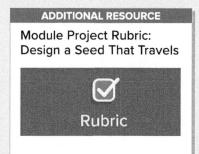

STEM Module Project   **Module: Plant Parents and Their Offspring / 87**

🖤 GO ONLINE

INTERACTIVE PRESENTATION	ADDITIONAL RESOURCE
Module Project: Design a Seed That Travels	Module Project Rubric: Design a Seed That Travels
Project	☑ Rubric

STEM Module Project
Engineering Challenge

Design a Seed That Travels

🕐 **Prep:** 5 min | **Class:** 60 min 👥 small groups

Read the paragraph in the student notebook aloud and explain the scope of the module project to the class. Explain that students will be helping a farmer spread seeds across his or her farm. Students will use engineering skills as they plan, build, test, and improve their design.

ASK: Why is it important that seeds travel? Sample answer: If seeds didn't travel, plants would grow in the same place as their parent and there would not be enough water and nutrients for all of the plants.

Project Parameters Explain to students that they will be designing and building model seeds to see how far they can travel. Encourage students to try to design the seed that will travel farthest.

Module Project Rubric Use the rubric to assess student understanding. Introduce the rubric before students begin to design their seed to ensure they know what is expected of them. Revisit the rubric as student designs progress to ensure all parameters of the project are met.

🖤 GO ONLINE to access the project rubric.

Build Your Model

Review lessons 1 and 2 with students. You may want to revisit the module vocabulary words or lesson phenomenon.

1. Have students look back at the lessons to determine three ways seeds travel.

MATH > Connection

Measurement and Data 1.MD.A.2

Remind students that they can use objects to make measurements. A distance can be measured by counting how many times an object (laid end to end) can fit within a certain space. When comparing lengths it is important students use the same object to provide a standard unit for comparison.

6. Have the class decide what object they will use to take their measurements. This object should be something easily available and standard across the classroom. Examples include: a book or an unsharpened pencil.

ASK: How will your seed travel? Sample answer: My seed will travel by the wind.

ASK: How far did your seed travel? Sample answer: My seed traveled four pencil lengths.

MODULE PROJECT

STEM Module Project
Engineering Challenge

Materials

It is recommended that you allow students to choose from a variety of different materials for this project. Examples include: chenille stems, tape, glue, hook and loop fasteners, construction paper, cotton balls, polystyrene foam, tissue paper, and toothpicks.

Design Your Solution

Help students create drawings of seeds. Have students label what material each portion of the light stand will be made of. Each group should circle the design they would like to build and have it approved by the teacher before moving forward. Remind students to refer to the rubric to make sure their model fits all the criteria.

Test Your Model

Have a designated area in the classroom or outside where students can test their seed models. Have students analyze and compare their results with other groups. Begin a class discussion where students identify which seed designs worked best and why. Give students the opportunity to make adjustments to their design and retest.

SEP Analyzing and Interpreting Data

Students should collect, record, and share their observations about their seed design with their classmates. It is important that students use the same object when measuring the distance their seed traveled, so the distances can be compared. Have students discuss which designs traveled the furthest and why those solutions were most successful.

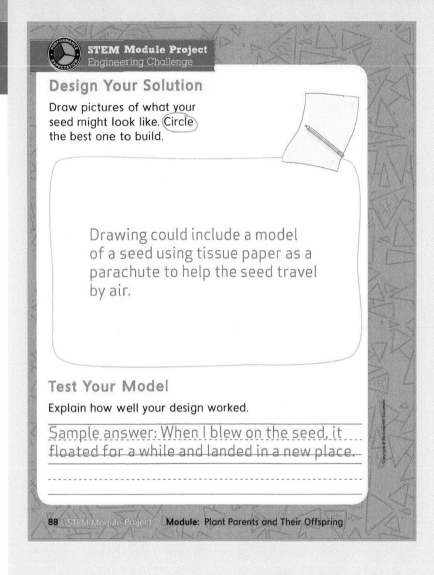

STEM Module Project
Engineering Challenge

Design Your Solution

Draw pictures of what your seed might look like. Circle the best one to build.

Drawing could include a model of a seed using tissue paper as a parachute to help the seed travel by air.

Test Your Model

Explain how well your design worked.

Sample answer: When I blew on the seed, it floated for a while and landed in a new place.

88　STEM Module Project　**Module:** Plant Parents and Their Offspring

Module Wrap-Up

REVISIT THE PHENOMENON

What happens when you blow on a dandelion? Draw a picture. Add labels.

Drawing could include a seed traveling by wind, water, or on an animal. Labels should include the structure that allows the seed to travel.

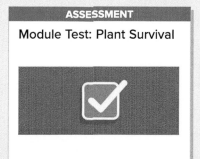

Look at your project to help you answer the question.

Module Wrap-Up **Module:** Plant Parents and Their Offspring **89**

MODULE WRAP-UP
REDISCOVER THE PHENOMENON

Look back at the module phenomenon showing a dandelion with seeds blowing in the wind. Have students discuss how seeds help the plant species survive.

Encourage students to relate this phenomenon to their module project.

ASK: How was your seed similar to or different from the dandelion seeds? Sample answer: I designed a seed similar to the dandelion seeds. My seed also used air to travel.

ASK: Can seeds decide where they want to travel? How do they move? No, seeds do not decide where to travel. Seeds travel wherever the wind, water, or animals take them.

GO ONLINE

INTERACTIVE PRESENTATION	ASSESSMENT
Module Wrap-Up: Plant Survival	Module Test: Plant Survival
WRAP-UP	✓

A

amphibian an animal that lives part of its life in water and part on land

B

behavior the way a person, animal, or thing acts or does something

bird an animal that has two legs, two wings, and feathers

C

communicate to give information about something

E

Earth the planet on which we live

energy the ability to do work

F

fall the season after summer

fish an animal that lives in the water and has gills and fins

flower plant part that makes seeds

fruit plant part that holds the seeds

function the purpose of something

H

horizon the line where the earth and sky seem to meet

I

illuminate to light up

inherit when something is passed from the parent to its young

insect an animal with three body sections and six legs

L

leaf plant part that makes food from sunlight

learn to gain knowledge or skill

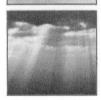

light a form of energy that lets you see

M

mammal an animal with hair or fur that takes care of and usually gives birth to live young

material what objects are made of

mirror a smooth surface that reflects what is in front of it

Moon a ball of rock that moves around Earth

Moon phases the different Moon shapes we see each month

N

need something you must have in order to live

O

 offspring young made by parents

 **opaque** materials that do not let light pass through

P

 parent a living thing that makes offspring

 pitch how high or low a sound is

 planet a very large object that moves around the Sun

 pollen powder that helps make new plants

 protection keeps things safe from harm

Copyright © McGraw-Hill Education (t to b)Flickr Open/Getty Images, Ken Karp/McGraw-Hill Education, Smith Photographers/Blend Images, McGraw-Hill Education, JPL-Caltech/Space Science Institute/NASA, McGraw-Hill Education, Sylvain Cordier/Getty Images

R

reflect give back an image

reptile an air-breathing animal that has dry skin covered with scales

root a plant part that keeps the plant in the ground, stores food, and absorbs water and nutrients

S

season one of the four parts of the year with different weather patterns

seed a part of a plant that can grow into a new plant

seedling a young plant

shadow a dark shape that is made when a source of light is blocked

signal a sound or movement that gives information

 sound a form of energy that comes from objects that vibrate

 spring the season after winter

 star an object in the sky that makes its own light

 stem plant part that holds up the plant

 **structure** a part of something

 **summer** the season after spring

 Sun the star closest to Earth

 **sunrise** the time of day when the Sun rises above the horizon

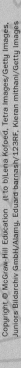

 sunset the time of day when the Sun descends below the horizon

 **survive** to live and grow

T

 **trait** a feature of a living thing

 translucent when some light can pass through

 transparent when light can pass through

V

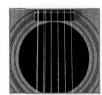

vibrate to move back and forth

volume how loud or soft a sound is

W

wave a movement up and down
or back and forth

winter the season after fall

Index

Folding Instructions

The following pages offer step-by-step instructions to make the Foldables study guides.

Half-Book

1. Fold a sheet of paper (8½" x 11") in half.

2. This book can be folded vertically like a hot dog or ...

3. ... it can be folded horizontally like a hamburger.

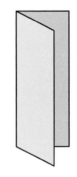
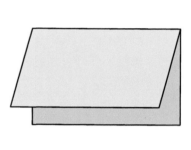

Folded Book

1. Make a Half-Book.

2. Fold in half again like a hamburger. This makes a ready-made cover and two small pages inside for recording information.

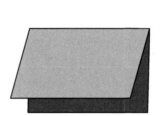

Pocket Book

1. Fold a sheet of paper (8½" x 11") in half like a hamburger.

2. Open the folded paper and fold one of the long sides up two inches to form a pocket. Refold along the hamburger fold so that the newly formed pockets are on the inside.

3. Glue the outer edges of the two-inch fold with a small amount of glue.

Shutter Fold

1. Begin as if you were going to make a hamburger, but instead of creasing the paper, pinch it to show the midpoint.

2. Fold the outer edges of the paper to meet at the pinch, or midpoint, forming a Shutter Fold.

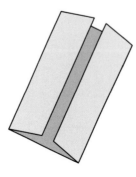

Trifold Book

1. Fold a sheet of paper (8½" x 11") into thirds.

2. Use this book as is, or cut into shapes.

Three-Tab Book

1. Fold a sheet of paper like a hot dog.

2. With the paper horizontal and the fold of the hot dog up, fold the right side toward the center, trying to cover one half of the paper..

3. Fold the left side over the right side to make a book with three folds.

4. Open the folded book. Place one hand between the two thicknesses of paper and cut up the two valleys on one side only. This will create three tabs.

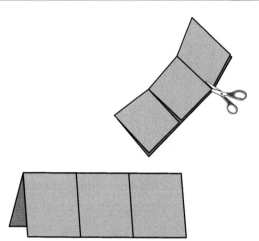

Layered-Look Book

1. Stack two sheets of paper (8½" x 11") so that the back sheet is one inch higher than the front sheet.

2. Bring the bottoms of both sheets upward and align the edges so that all of the layers or tabs are the same distance apart.

3. When all the tabs are an equal distance apart, fold the papers and crease well.

4. Open the papers and glue them together along the valley, or inner center fold, or staple them along the mountain.

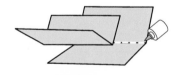

Folded Table or Chart

1. Fold the number of vertical columns needed to make the table or chart.

2. Fold the horizontal rows needed to make the table or chart.

3. Label the rows and columns.

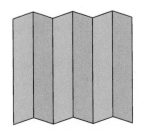

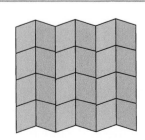

An Interview with

Dinah Zike Explaining
Visual Kinesthetic Vocabulary®, or VKVs®

What are VKVs and who needs them?

VKVs are flashcards that animate words by kinesthetically focusing on their structure, use, and meaning. VKVs are beneficial not only to students learning the specialized vocabulary of a content area, but also to students learning the vocabulary of a second language.

Dinah Zike | Educational Consultant
Dinah-Might Activities, Inc. – San Antonio, Texas

Why did you invent VKVs?

❝ Twenty years ago, I began designing flashcards that would accomplish the same thing with academic vocabulary and cognates that Foldables® do with general information, concepts, and ideas—make them a visual, kinesthetic, and memorable experience. ❞

I had three goals in mind:

Dinah Zike's
Visual
Kinesthetic
Vocabulary®

- **Making two-dimensional flashcards three-dimensional**

- **Designing flashcards that allow one or more parts of a word or phrase to be manipulated and changed to form numerous terms based upon a commonality**

- **Using one sheet or strip of paper to make purposefully shaped flashcards that were neither glued nor stapled, but could be folded to the same height, making them easy to stack and store**

Why are VKVs important in today's classroom?

❝ At the beginning of this century, research and reports indicated the importance of vocabulary to overall academic achievement. This research resulted in a more comprehensive teaching of academic vocabulary and a focus on the use of cognates to help students learn a second language. Teachers know the importance of using a variety of strategies to teach vocabulary to a diverse population of students. VKVs function as one of those strategies. ❞

How are VKVs used to teach content vocabulary to EL students?

““ VKVs can be used to show the similarities between cognates in Spanish and English. For example, by folding and unfolding specially designed VKVs, students can experience English terms in one color and Spanish in a second color on the same flashcard while noting the similarities in their roots. ””

What organization and usage hints would you give teachers using VKVs?

““ Cut off the flap of a 6" x 9" envelope and slightly widen the envelope's opening by cutting away a shallow V or half circle on one side only. Glue the non-cut side of the envelope into the front or back of student workbooks or journals. VKVs can be stored in the pocket.

Encourage students to individualize their flashcards by writing notes, sketching diagrams, recording examples, forming plurals (radius: radii or radiuses), and noting when the math terms presented are homophones (sine/sign) or contain root words or combining forms (kilo-, milli-, tri-).

As students make and use the flashcards included in this text, they will learn how to design their own VKVs. Provide time for students to design, create, and share their flashcards with classmates. ””

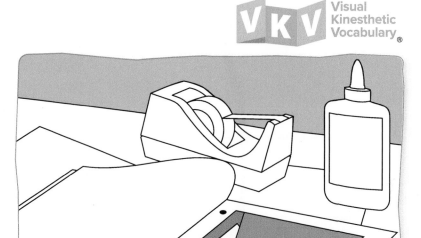

Dinah Zike's
Visual
Kinesthetic
Vocabulary ®

Dinah Zike's book Foldables, Notebook Foldables, & VKVs for Spelling and Vocabulary 4th-12th won a Teachers' Choice Award in 2011 for "instructional value, ease of use, quality, and innovation"; it has become a popular methods resource for teaching and learning vocabulary.

Compare and Contrast

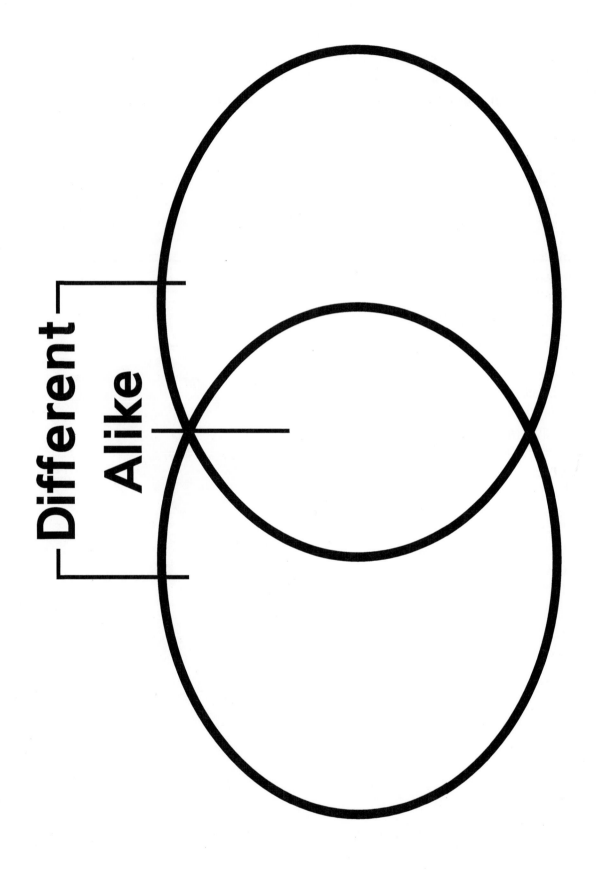

Problem and Solution

Problem

Steps to Solution

Solution

Name _____ Date _____

Sequence

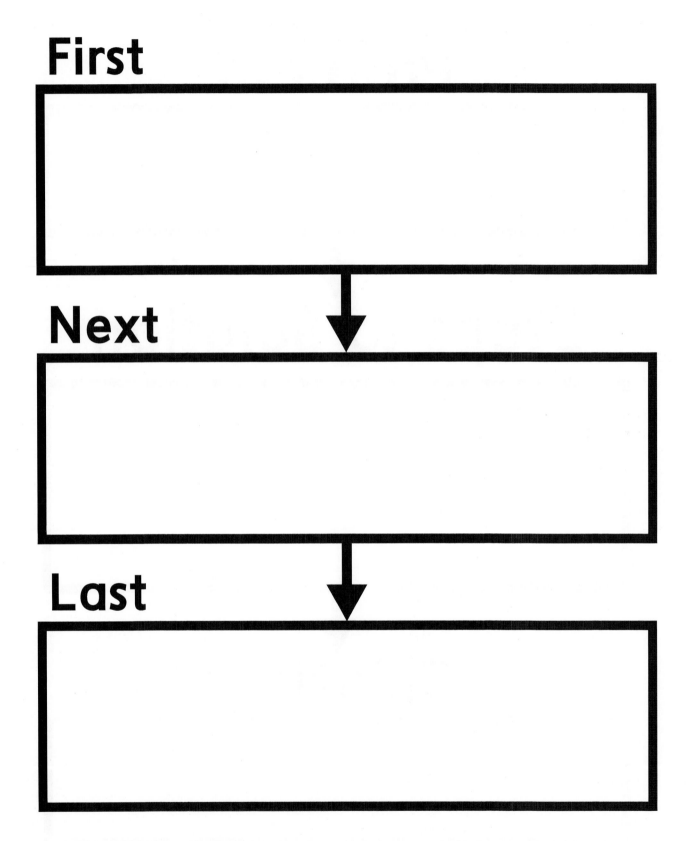

First

Next

Last

Classify

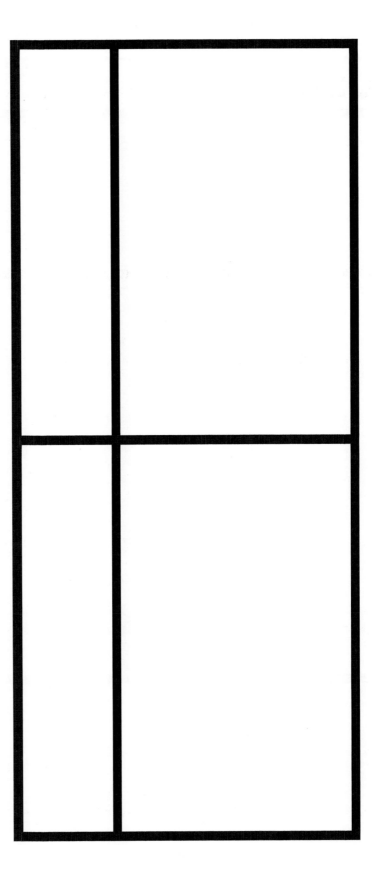

Cause and Effect

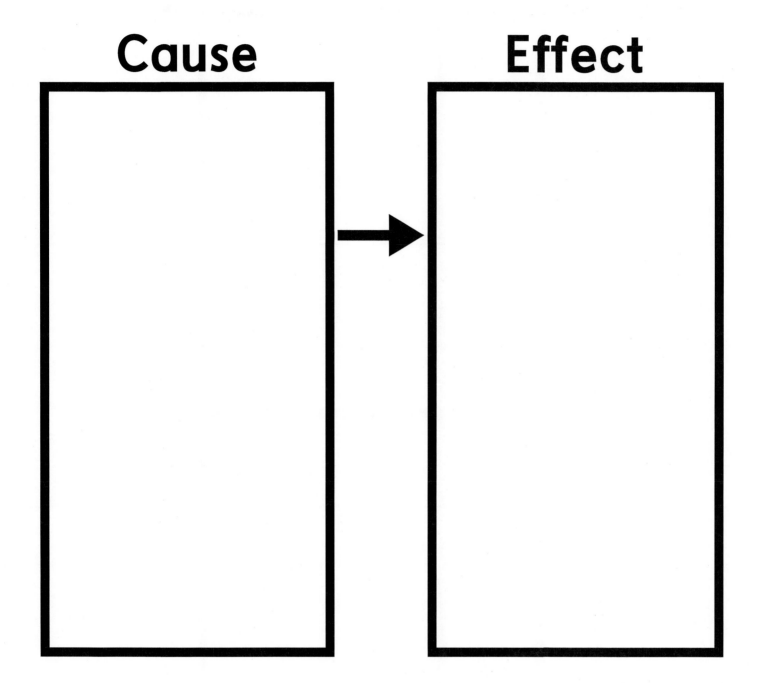

Cause

Effect

Infer

What I Infer	What I Know	Clues

Name _____ Date _____

Draw Conclusions

Text Clues	Conclusions

Summarize

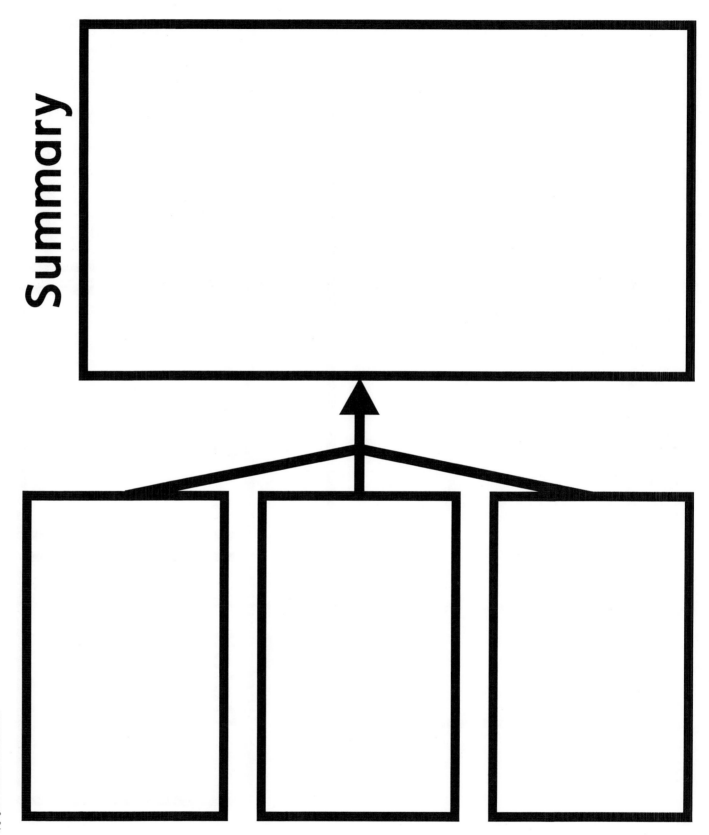

Main Idea and Details

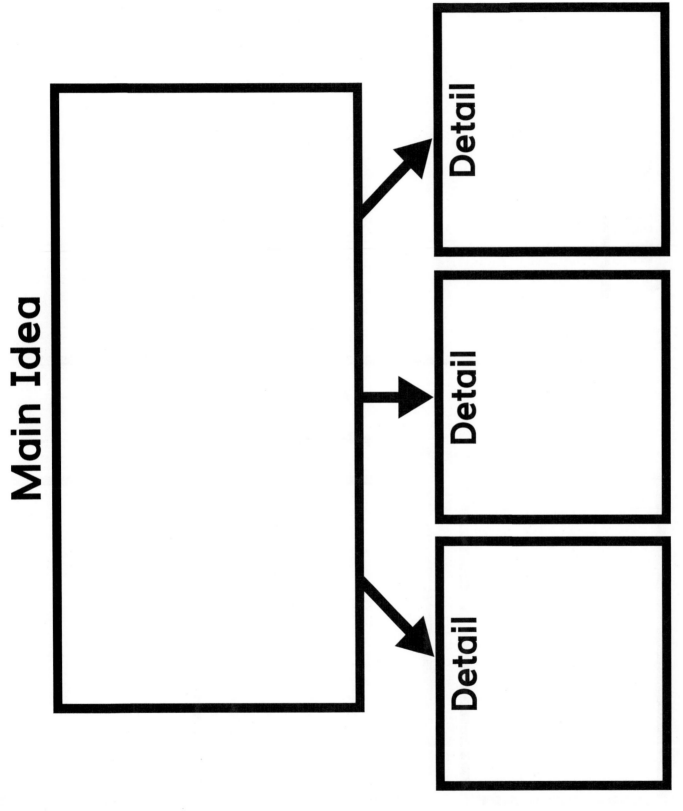

Main Idea

Detail

Detail

Detail

Notes